Dictionary of Fire Protection Engineering

Clifford Jones

Whittles Publishing

Published by
Whittles Publishing,
Dunbeath,
Caithness KW6 6EY,
Scotland, UK

www.whittlespublishing.com

© 2010 J.C. Jones
ISBN 978-1904445-86-9

Printed and bound in England
www.printondemand-worldwide.com

Contents

Dedicated to

Professor David Trimm AM

Colleague of the author's over the period 1987–1995

Preface

In writing this book I have tried to hold in balance a number of areas of interest to the fire engineer. As a reader would expect, such matters as fire extinguishers and flame retardants feature fairly centrally. There is also a good deal about fire fighting and this includes descriptions, from the functional point of view, of fire appliances from selected manufacturers around the world. There is coverage of selected accidental fires, both recent ones and ones which have been on record for many years as being amongst the most serious in terms of loss of life. In discussing a fire which is distant in time I have tried to combine the facts as far as they are reliably known with professional judgement of my own. There have been times when such judgment has put me in some degree at odds with one or more details of a fire as recorded. In such cases I have occasionally exercised the prerogative of the expert in analysing the events for myself and presenting a reader with conclusions of my own. Social and political aspects of fire engineering also feature in this book, for example in accounts of fires in countries where buildings are substandard in safety terms and fire services are unreliable.

It is inevitable that trade names have been used. Avoidance of them would have impoverished the book, as fire safety products are so much a part of the subject and improvements in fire safety have to a considerable degree been due to development work by manufacturers of such products. My own position in using a trade name is of course an entirely neutral one, and the aim has been to obtain scientific and engineering details of the products and re-express such details in broad terms for the benefit of the reader. There is a great deal of extremely interesting and helpful material in the trade literature. It was quite early on in my own career, long before the "IT revolution", that I became aware that a well composed technical manual accompanying a piece of equipment can in fact make a very rewarding read.

Almost all of the sources I have drawn on have been electronic, and my indebtedness to the originators of those sources is immense. Throughout the writing of the book I have had the encouragement and support of Whittles Publishing with whom I have by now had a close working relationship for over a decade.

J.C. Jones
Aberdeen

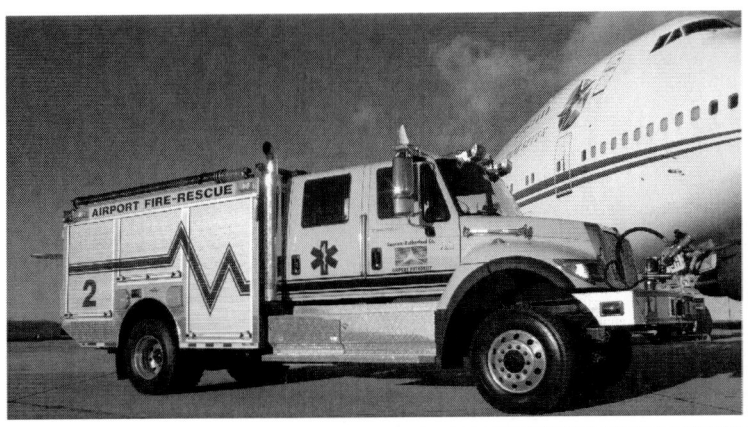

Rosenbauer Class 3 Airwolf (courtesy Rosenbauer/General Division)

A

Aarhus

This Danish town was the scene in late 2007 of a fire at a factory where vegetable oils are processed for subsequent use in industries including chocolate and cosmetics. One employee was killed and damage was considerable.

ABC powder extinguisher

Substance suitable for extinguishment of Class A (burning solids, such as wood, polymers), Class B (liquid fires, such as alcohol, kerosene) and Class C (fires involving flammable gases, such as natural gas, propane) fires with the possible bonus of also being suitable for electrical fires (Class E). Examples include **monoammonium phosphate**. Nitrogen under pressure acts as a delivery gas, enabling the powder to be rapidly released and directed as required. A "BC fire extinguisher" is suitable for Class B and Class C fires only: by far the most important example is the **carbon dioxide extinguisher**. In a powder extinguisher this is obtained by release of **sodium bicarbonate** or **potassium bicarbonate,** which decompose to form carbon dioxide.

Abeille Bourbon

French coastguard vessel built in 2005, having been designed by Rolls Royce. It has **external fire fighting** facilities at **FiFi II** level, there being three monitors of 2400 m³ per hour capacity and a throw of 150 m.

Abeille Liberté

Sister vessel (younger by a few months) of **Abeille Bourbon**. Lloyd's List[1] shows an image of the vessel releasing water from its monitors, captioned: "*Abeille Liberté* demonstrates its awesome fire fighting capability using high standard equipment". This provides a perspective on the requirements for

---

1   17 March 2006.

FiFi II classification. Many tugs newly commissioned for work in conjunction with refineries, including the **Stanford**, have **FiFi I** classification only.

Abesco CT120

Integrally designed duct for cables incorporating a graphite-based intumescent material. The action of this in a fire seals off the duct, preventing combustion propagation along it. The duct can be installed in concrete ceilings and floors, and also in wood structures provided that standards and codes that apply to the wood *per se* have been met.

Acergy Harrier

Construction support vessel for offshore installations. Her function is quite different from that of an AHTS vessel, but she shares with most such fire fighting capability at **FiFi I** level.

Active Lord

Supply vessel in the North Sea, an example of a vessel with dual FiFi classification being I and II. Water supply is up to 7200 m³ per hour via three **monitors**. **Atlantic Eagle, Bonassola** and **Stril Poseidon** are similarly classified.

Adjustable gallonage nozzle

In the **fixed gallonage nozzle** the gpm–pressure data pair is determined by the diameter of the orifice (flow disc). In an adjustable gallonage nozzle changes in the orifice diameter are possible, providing for a change in gpm without interruption to supply. The gallonage is therefore incrementally adjustable, not continuously so. That is why the term "selectable flow nozzle" is sometimes preferred. Nozzles of this type are usually fitted with "teeth" so as to issue water as a fog rather than as a straight stream. Examples are very many and include the **ThunderFog 250**.

Advanced pneumatic detector (APD)

Pneumatic detector, utilising a metal hydride, for detection of overheating in a jet engine during flight. It is now manufactured by Kidde Aerospace and Defense in North Carolina. Note that whereas the basis of thermocouples and and thermistors is electrical that of a device such as the APD is *mechanical*.

Advantus™

Foam proportioner, developed and manufactured by **Waterous** and the first in a new generation of proportioners. Unlike other types of proportioner, including the **eductor**, Advantus™ does not combine the water and foam concentrate on a volumetric basis but continually monitors both water and concentrate compositions and proportions on that basis.

Aerial ladders, materials for

Raised and orientated by hydraulic machinery, an aerial ladder, unlike a ground ladder or a roof ladder, is not supported by anything external to itself such as a wall. This precludes adaptation of the **fibre glass ladder** to aerial usage. An aerial ladder has to support its own weight (i.e. it must not bend significantly) as well as hose along its length. A platform ("**ladder bucket**") might be attached: trade terminology in the US is that the appliance is "an aerial" if a ladder bucket is not fitted and "a platform" if it is. Either might be classifiable as a **quint** if it is has the additional facilities for such classification. Materials used for aerial ladders are *aluminium* or *steel*. Water will be directed via **monitors** on the platform and some manufacturers install lights along the ladder rails. Hose chosen for use with aerial ladders will be light, for example **DarQuest Lightweight** or LDH (from the manufacturer of **Pro-Lite**). The weight which can be withstood by a ladder or a ladder/platform combination depends on the flow speed of any water in hoses being supported. The fire fighting vehicle supporting the aerial ladder will itself need supporting by means of stabiliser jacks after the ladders have been raised.

Aero Range

Non-maintained emergency lighting, powered by a battery from which it draws 8 W yielding 70 lumen, i.e. a yield of just under 9 lumen per watt. The transparent cover is made of polycarbonate.

Aerospatiale SA-315B ("Lama")

Helicopter suitable for use in fire fighting when fitted with a **Bambi bucket®** and related accessories. One "Lama" in fire fighting duty crashed in the Rockies in 2002, killing the pilot. The helicopter was 32 years old at the time of the crash. Several remain in fire fighting use in the US.

AFFF

Acronym for "aqueous film-forming foams". These are made from water, a foaming compound and a surfactant. An AFFF is made by mixing water with a suitable concentrate, e.g. **Tridol S 3LT**. There are many compositions of film-forming concentrates and the term AFFF is fairly generic, there being acronyms for particular foam concentrates which give more information than "AFFF" does. It would not be incorrect, though it might be imprecise, to call a foam concentrate of the **FFFP** type an AFFF. The acronym AR-AFFF applies to **Ultraguard**. Common ingredients of AFFF have included **perfluorooctanoic acid** and **perfluorooctane sulphonate**.

AFG Firewall™

Gel concentrate similar to **Thermo-Gel®** in its uses but containing only one of the polymeric constituents (polyacrylamide) of Thermo-Gel®. A Tyco product, it finds wide application to forest fires when it is combined with the extinguishing water by means of an **eductor** at levels of 3–4%.

African townships, fire hazards in

Townships exemplified by Soweto near Johannesburg proliferated in South Africa during the apartheid years and have not yet disappeared. Amongst their many shortcomings is the fact that they are a fire hazard and there are more fire calls to such townships in the Transvaal than to downtown Johannesburg where, until the dismantling of apartheid, natives were not allowed to live. Adding to the dangers due to wooden construction of dwellings is the fact that many of them contain kerosene heaters in very poor condition.

Air bags, possible behaviour of in a car fire

An airbag can come into operation in a vehicle fire long after initiation, during emergency responses. This can endanger rescue workers. A difficulty is that it is widely believed that to disconnect the car battery is to preclude airbag deployment. This is not true as the air bag's opening device can store some charge, and if the air bag comes into operation whilst a rescue worker is extricating an occupant of the car it can trap both. Rescue workers are therefore trained to treat any airbag which has not operated as being capable of doing so.

AirBoss

Widely used series of **SCBA** manufactured by Draeger. Like some other such products, it has an air supply that is "first breath activated", that is it commences the first time the wearer attempts to inhale. It can be used with an aluminium cylinder or with one made from a composite containing **carbon fibre**. The weight advantage of such composites is evident when specifications for the AirBoss are examined, there being a difference of 4.4 kg between a full aluminium cylinder at 2216 psi and a full "carbon composite" one filled to the same pressure. "**Buddy breathing**" is an optional extra with this apparatus.

Air conditioners in cars, vulnerability of

Being positioned close to the leading edge of a car body so as to experience good forced convection, a car air conditioner is likely to break open and release its contents in even a quite a mild collision. By contrast the gasoline tank is protected from rupture by the body panels. In the move to flammable fluids such as **HC-12a®** for car air conditioners this is a factor. The oft-repeated argument that "a car carries flammable fuel and brake fluid, so why can't it carry flammable air conditioner fluid?" carries very little weight. (See also **Panama, bus fire 2006**.)

Aircraft fuel pumps, chafing of wire in

Many jets in the US were examined for such chafing after it came to light that with a particular make of pump the wires were too close to the rotor. It was emphasised that no incidents had ensued from this possible fault. It was also pointed out that chafed wires would not lead to an ignition as long as the pump was totally immersed in the fuel liquid and that there was only a risk if the pump was in contact with vapour. In general in aviation use, wire chafing is exacerbated by vibrations.

Aircraft fuel tanks, fire hazards with

Aircraft fuel tanks are vented to the atmosphere, meaning that the space above the liquid surface is occupied by a fuel vapour/air mixture at the pressure corresponding to the altitude. Jet fuels approved for use by the FAA tend to have pressure–temperature characteristics such that at a total pressure of 1 bar below 38°C the vapour–air mixture in the tank is too lean to ignite and above 80°C it is too rich to ignite. At cruising altitude the total pressure will of course be well below 1 bar. These facts indicate that at cruising height fire in the tanks is precluded but not necessarily during take-off and landing. It is in fact understood that for a small proportion of its time in the air a jet will have in its tanks an ignitable vapour–air mixture. This was stated in the follow-up to a fire involving a TWA 747 aircraft taking off from JFK Airport in New York in 1996. Many US aircraft were lost during the Vietnam War through explosion of the fuel tank due to enemy ground fire, and means of inerting the tanks were developed. The inerting materials are nitrogen, which might be carried as liquid nitrogen, and carbon dioxide carried as dry ice. Inerting comes into operation as required by the pilot and is not continuous as the quantity of inerting material carried provides for protection only when under fire. The situation with *commercial* aircraft is that since the 1996 TWA crash the FAA are starting to require inerting. Onboard inert gas generator systems (OIGGS) are available whereby air passes through a membrane structure which removes some of the oxygen, a mixture down to 12% oxygen balance nitrogen being obtainable by this means. Many US airlines are at the present time struggling for financial survival, and a positive injunction to retrofit aircraft with inerting devices could drive some of them out of business. (*See also* **Boeing 737, two incidents with**; **Lockheed C-130K**.)

Aircraft windshield heaters, as an ignition source

An electrical terminal block in the windshield heater is believed to have caused two fires in Boeing 757 aircraft, both in 2004. One of them was taking off from Dallas/Forth Worth Airport at the time of the incident and the other was on the tarmac at Copenhagen Airport. In neither case were there serious consequences. Boeing stated that the same difficulty had been reported with four other aircraft and that a redesigned terminal block was being fitted to aircraft manufactured after mid 2004.

Air cylinders for fire fighters, reconditioning of

In the US air cylinders for **SCBA** conform to ASME[2] or DOT[3] design standards. There is a huge turnover of such cylinders ("bottles" as they are often called) and major manufacturers of breathing apparatus including Scott (in upstate New York) and Dalmatian (in Utah) buy cylinders from fire departments and recondition and "requalify" them. A cylinder will of course by that time have undergone many refillings by the department which previously owned it. "Visual condition inspection" of a cylinder for reuse begins with noting the serial number. A cylinder the serial number of which is illegible is not suitable for reconditioning. There should also be a record, on the cylinder surface itself, of the initial pressure testing and of any more recent pressure testing. These data – serial number and pressure test records – give a cylinder an "identity" and this is an important quality control factor. Further visual inspection is of the surfaces for corrosion and damage. Minor damage or corrosion does not necessarily preclude reuse, and criteria apply. The condition of the thread from the cylinder is also a factor; rethreading is sometimes possible. The mating connector bearing attachments has an O-ring at its base and it is usual for this to be replaced when a cylinder is reconditioned. Cylinders have been kept in circulation for very many years in this way. In the US there is a maximum of 15 years imposed on the use of a cylinder made from a composite containing a **carbon fibre**. Other components of **SCBA** can also be purchased in reconditioned form.

---

2   American Society of Mechanical Engineers.

3   Department of Transport.

Air lifting bag

When used in fire fighting to lift structures or to remove heavy objects restricting access, these are inflated with compressed air. **Aramid fibre** is sometimes used in their manufacture. Examples include the **Maxiforce™** range.

Air tanker

A.k.a. (especially in North America) as a water bomber, fixed-wing aircraft for aerial fire fighting. It will often be escorted to its destination by a **lead plane**. An example of an air tanker currently in service is the **Tanker 910**. At the other end of the size range is the **Dromadear**. The earliest, based on a Boeing Stearman biplane, departed **Willows, CA** for its first voyage in fire fighting duty in August 1955.

Airwolf

Newly introduced **ARFF** vehicle produced by Rosenbauer, a single one of which would satisfy the requirements for a Category 3 airport according to the **NFPA categories of airports**. Its water pumping capability is 500 gpm., and it carries foam as well as dry chemical agent. It is the fourth ARFF vehicle to have been built by Rosenbauer. The others are named Panther, Buffalo and Simba. The Panther is on a bigger scale than the Airwolf and can supply extinguishment water at up to 1800 gpm. A new Panther was recently delivered to Dubuque Regional Airport, Iowa and another to the Irish Air Corps in Dublin. The Simba was to some extent custom built to meet the particular needs of Frankfurt Airport.

AJ-Elite A

Oil for use in industrial mechanical pumps, a synthetic substance having a flash point as high as 249°C. The sister product **AJ-Elite Z** has a similar flash point but is of higher viscosity.

Al-Ahmadi Refinery, fire in 2000

This refinery is close to the Burgan oil field, from which it receives crude, and is about 20 miles from Kuwait City. A fire there caused by leakage of natural gas in June 2000 killed two persons and injured 49 others.

Alarm number

The fire at **East Orange, NJ** was described in the press coverage from which the information was taken as a "two-alarm fire". This denotes the number of alarms communicated to the fire department members after information in the call has been assessed, the more serious the fire the greater the number of alarms and the more crew and fire trucks mobilised. That much is fairly standard across the US although, as would be expected, fire departments differ in the scale of responses. This is partly because of different sizes and capitilisation with equipment: the FDNY would send to a fire at the low extreme of the alarm number scale more fire trucks than some fire departments would possess. Moreover the alarm number assigned at initial response can be raised if more help is needed, or lowered if some appliances and their crews can be returned to base. The FDNY sends to a two-alarm fire 25 "units" and 106 fire fighters. "Unit" means "vehicle" and this can vary in size from an aerial ladder to a sedan car driven by a fire chief. A three-alarm fire receives 33 units and 138 fire fighters. Continuing to use the FDNY as an example, the May 2006 fire at the Brooklyn water front, which 90 appliances and fire fighters attended, was initially categorised as eight-alarm, later revised to ten-alarm. It should be noted that the numeral is in no way a multiplying factor: a ten-alarm fire does not require five times the fire fighters and appliances that a two-alarm one does. (See also **Bronx, apartment fire 2006**; **Greenwich Village, apartment block fire 2004**.)

Alaska paper birch

Botanical name *Betula neoalaskana*. This species of tree occurs from Alaska across to Newfoundland. The potential of a particular tree as a **firewise plant** might depend on the location, its growth behaviour being influenced by subtle changes to the ecosystem from one location to another. Within Alaska the Alaska paper birch is recommended as being firewise but not necessarily elsewhere. This is because under the conditions in Alaska it produces debris ("browse") only to a very small degree but will do so more abundantly in certain other regions. Having a thin bark, the tree structure itself is quite susceptible to fire.

Albi Clad TF

Water-based intumescent material containing chlorinated organics and a titanium compound. It has found wide application at the outsides of airport terminals. It is sometimes used to coat steel structural members, where its action in the event of fire is to prolong the time over which the steel can receive heat from a fire without failing in its support role.

Alexis

Fire truck manufacturer, based in the Illinois town of the same name. It builds trucks across the range of types and also offers refurbished fire trucks for sale, some of which first entered service decades ago. Amongst its recent deliveries of new trucks is a small (overall length is

2.9 m) **rescue vehicle** on a Chevrolet chassis and cab structure having a 230 hp Caterpillar diesel engine. Its payload includes illumination facilities capable of 6000 W. Of the "non walk-in type", it cannot carry human "cargo". A somewhat larger **rescue vehicle** recently supplied by Alexis (to Wapello, IA) *is* "walk-in". Of overall length just under 9 m, it has a Caterpillar 330 hp diesel engine and power take-off (PTO) electricity generation, the supply from which can be used to provide lighting. (*See also* **Joy, IL**; **Peoria heights, IL**.)

Algeciras Carrier

Cargo vessel bearing the flag of the Bahamas, detained at a UK port in circumstances similar to those of the ***Berkan B***. As well as fire damper deficiency, the *Algeciras Carrier* had many sites of leaked oil posing an ignition hazard.

Allied Colloids, Bradford, fire 1992

This fire occurred when AZDN (azodiisobutyronitrile) and SPS (sodium pyridine-3-sulphonate) came into accidental contact. There were no deaths or injuries. The most serious consequences were contamination of the Aire and Calder rivers with fire water. Since 1998 Allied Colloids has been part of the Ciba group.

AllRounder

Fire appliance manufactured by **Angloco**, having an aerial ladder and water storage and pumping capability. It might therefore be seen as a **quint** although there is a subtle difference between it and aerials so classed in the US. In the latter the appliance is configured at design stage as an aerial with the ladder turntable either at the rear or at the centre and other facilities are supplementary. The AllRounder has the basic configuration of a water tanker with a turntable mounted immediately behind the cab.

Alpha-olefin sulphonate

Agent used in the production of foam for fire fighting ("fire suppressant foam"). An alpha-olefin sulphonate solution is the major constituent of Phos-Check® WD881, from the same manufacturer as **Phos-Check® AquaGel**, as it is of Stepantan® AS-12 46, manufactured by the Stepan Company whose HQ are in Illinois.

Aluminised fire apparel

Materials such as **PBI™**, **Nomex®** and other fabrics suitable for protection of fire fighters can be coated with aluminium. Coats, jackets and gloves can be so treated and their advantage is the capability of the surface to reflect heat. A note on the relevant physics is helpful here. Visible light encompasses only a small part of the thermal radiation wavelength range. A surface which reflects in the visible *might* be strongly absorbing in other regions of the spectrum; in the terminology of radiation heat transfer that means that the surface is "non-grey". So reliance on reflection of the narrow range of wavelengths which the human eye can see is unsound. The author is in no way deprecating the use of aluminised fire apparel, having no background in R&D which the respective manufacturers have carried out. He is simply encouraging readers not to slip into the error of supposing that something reflecting in the visible is necessarily good protection from thermal radiation.

6061 Aluminium

Alloy having >95% aluminium, often used with aluminium itself in the construction of plant, including aerial ladders and the **ladder bucket**, also in panel work for a fire truck itself. Other elements present include silicon, iron and copper, all in very small amounts. In more casual discussions and reporting "aluminium" might mean 6061 aluminium. The term "extruded aluminium" sometimes means 6061 aluminium.

Aluminium trihydroxide (ATH), use of as a fire retardant

In discussing this substance we first note that:

$$2Al(OH)_3 \equiv Al_2O_3.3H_2O$$

hence the synonymous use of the terms aluminium trihydroxide and alumina trihydrate, each of which fits the acronym ATH. The second term is more accurate in that without doubt one of the ways in which ATH suppresses fire is by absorbing the heat required for the process:

$$Al_2O_3.3H_2O \rightarrow Al_2O_3 + 3H_2O$$

In a substance such as a polymer, the cooling so afforded will decelerate pyrolysis reactions which provide the flammable gaseous materials required to sustain a flame. ATH can be manufactured from bauxite via sodium aluminate, obtained by treating the bauxite with caustic soda. Annual production of ATH exceeds 20 million tonnes and materials to which it is applied as a fire retardant include rubber, electrical cables and carpets. There are many proprietary fire retardants containing ATH the performance of which will depend, for example, on the particle size. An example is **FlameGard®**.

Aluminium wiring

Because of the soaring price of copper, about 40 years ago there was a move in countries including the US, the UK and Australia to substitute aluminium for copper in the provision of electrical power to buildings. Aluminium conductors *per se* are not dangerous. Difficulties can however occur when an aluminium cable is taken to a connector of different composition, perhaps steel. Supply of current and cessation of supply as appliances are switched on and off lead respectively to heating and cooling of the conductors, and aluminium and steel respond differently in their expansion and contraction behaviour. This can lead to loosening of the connection and possibly to eventual formation of a fused blob of metal which can act as a heating element. Where such problems develop (and are found during a mandatory periodic inspection) replacement of the wiring is not necessary, only replacement of connectors with ones made of materials known to be suitable for use with aluminium. Insurance companies have been known to require that this measure be taken.

American Darling

In the US fire hydrants are above ground and hose is connected to them as required. The American Darling series of hydrants owe their name to the fact that the original manufacturer was the Darling Valve and Manufacturing Company, who entered the business in 1902. American Darling hydrants are still in widespread use over a hundred years later, being manufactured by American Valve and Hydrant, Beaumont, TX. There are several series of American Darling hydrants being produced at the present time. As an example, the American Darling 84-B-84 series has at the base where the hydrant contacts the water supply a valve of 5.25 inch diameter and this is not intended to control the flow rate: the valve is either opened or closed and will be at a burial depth of typically 6 feet. The cast iron barrel through which water passes after valve exit is designed for a working pressure of 250 psi. The 84-B-84 1291 model has two 2.5 inch outlets for connection to hose and one 4.5 inch outlet. The 84-B-84 2400 model has just two 2.5 inch outlets. Each is equipped with a traffic flange, by means of which in the event of severe vehicle impact the "sacrificial bolts" at the base fail and the hydrant itself is spared damage. American Darling hydrants are of the dry barrel type, that is the hydrant enclosure (barrel) is empty when the hydrant is out of use.

American LaFrance, recent deliveries by

American LaFrance, based in South Carolina, have been one of North America's leading manufacturers of fire appliances for as long as such appliances have been in existence. Like many other such manufacturers, they display their recent deliveries on their web pages. The table on the next page is a selection by the author of details of 13 "recently delivered" appliances from American LaFrance. The entries are in ascending order of appliance size.

The appliances vary in size from FSD-operated ambulances to a **tractor-drawn aerial** and the vehicle length increases by a factor of 2.3 in going from the former to the latter. In going from two to three axles an increase in power of about 100 hp is evident: comparison of the aerial delivered to Danville, PA and those delivered to Newburgh, NY and Lakeville, MN shows this trend. The **tractor-drawn aerial** shares an engine with the three-axle aerials which are not tractor drawn. The tractor drawn aerial does not have a water tank or a pump, and this is the norm. As noted in a subsequent entry, the Detroit Diesel 515 hp unit is used by other builders of fire trucks including **Sutphen**. The places of delivery of the appliances in the table range from Florida in the south east to British Columbia in the north west. Supply is through regional dealers who also provide support after a sale. American LaFrance fire trucks are in use outside North America at places including Queensland, Australia. (*See also* **Fire appliances, used and rebuilt**; **Rescue vehicle**.)

Americium, use of in smoke detectors

An ionisation smoke detector requires ionising radiation, and this is provided by inclusion of a quantity of the order of a microgram of the isotope ^{241}Am into such a device. The americium emits alpha particles and gamma rays: the latter are produced at a level too low to be a hazard to persons, and exit the smoke detector device. The former stay within it and, by impingement with oxygen and nitrogen molecules in air, cause ion formation with an associated current. Smoke in the air passing into the device will absorb the alpha particles, leading to a drop in ionisation and of the current, which is the basis of an emergency signal.

Ammonia, fire hazards with

Ammonia is of course flammable, having a calorific value about a third that of natural gas. Its uses as a chemical include refrigeration, when, in particular at the compression stage of the cycle, leakage can occur leading to a fire hazard as at **Cartersville, GA**.

Ammonium polyphosphates

General formula $(NH_4)_k H_{(n + 2 - k)} P_n O_{(3n + 1)}$, the number of monomeric units rarely exceeding 1000. Ammonium polyphosphates are often a component of fire retardants which display **intumescence**. Commerical retardants

13 "recently delivered" appliances from American LaFrance

Place of delivery	Axles	Equipment	Length/m	Power unit	Livery
Dania Beach, FL	Two	Paramedic requisites	7.6	Mercedes Benz 210 hp	Red
Falmouth, MA	Two	Paramedic requisites	7.6	Mercedes Benz 210 hp	Red
Laredo, TX	Two	Pump Tank (1000 gallon)	8.0	Mercedes Benz 330 hp	Red
Piercefield, NY	Two	Pump (1250 gpm) Tank (1800 gallon)	8.0	Mercedes Benz 330 hp	Pale green
Eagle Bay, BC	Two	Tank only (2000 gallon)	7.2	Mercedes Benz 300 hp	Red
Castleton, NY	Two	Pump (1500 gpm) Tank (1000 gallon)	10.7	Cummins 400 hp	Blue

Table continued

Danville, PA	Two	Turntable ladder (65 feet reach) Pump (2000 gpm) Tank (500 gallon)	10.0	Detroit Diesel 455 hp	Red and white
Greenville, IL	Three	Pump (2000 gpm) Tank (2500 gallon)	11.9	Detroit Diesel 515 hp	Red and white
Ringwood, NJ	Three	Pump (1750 gpm) Tank (2500 gallon)	11.3	Detroit Diesel 515 hp	Red and black
Newburgh, NY	Three	Turntable ladder (100 feet reach) Pump (2000 gpm) Tank (300 gallon)	14.8	Detroit Diesel 515 hp	Pale green and white
Lakeville, MN	Three	Turntable ladder (100 feet reach) Pump (2000 gpm) Tank (300 gallon)	14.0	Detroit Diesel 515 hp	Red
Roseville, CA	Three, in a tractor-drawn arrangement	Turntable ladder	17.2	Detroit Diesel 515 hp	Red and white

containing ammonium polyphosphates include **Exolit AP**, **Antiblaze®** LR2 and Hostaflam AP.

Anaga, Tenerife

The scene in July 2006 of an accident involving a **Sikorsky S61** helicopter in fire fighting duty which claimed the lives of all six crew. The helicopter had been chartered from the Spanish concern Helicsa whose fleet of S61s have also seen service at offshore installations off various countries including Norway.

Anglegarth

Anchor handling tug supply (AHTS) vessel built in 1996. It is in use in UK waters, being equipped additionally as a fire fighting vessel at **FiFi I** level, having two **monitors** which have different capacities. The port side one can deliver 1200 m^3 per hour of water and the starboard side one 1500 m^3 per hour. That is because the starboard side monitor doubles up as a "drench system", some of the water it releases being used to protect the vessel itself. The vessel also carries foam. Amongst the emergencies to which the *Anglegarth* has responded was that involving the vessel *Autopremier* in 2001.

Angloco

Manufacturer of fire appliances, based in Batley, West Yorkshire. It sources cab/chassis units from suppliers including Mercedes, Ford, Daf, **Scania**[4] and (for small appliances) Land Rover. An order for a new apparatus can be discussed from the initial design stage with the fire brigade to which it is to be supplied. Amongst its recent deliveries is a tender of 475 gallon capacity on a Scania chassis with pump and hose. Angloco has had export orders from places including **St Lucia**.

Angola, fire services in

Angola is the only country in southern Africa with major oil and natural gas production, there being such production on- and offshore. Of importance not only to Angola but to the entire world was admission of Angola to OPEC in 2007. This of course involves maintenance of a quota of oil production set by the OPEC conference and signifies development of the country. For a few years before admittance to OPEC Angola was improving her fire service,

---

4   For an impressive shot of a fire truck bearing both Angloco and Scania badges, in service in Hong Kong, go to: http://www.fire-engine-photos.com/picture/number2824.asp

by creation of new infrastructure, by acquisition of new fire fighting plant, including a number of boats, and by training for fire fighters. Like Nigeria (also in OPEC) Angola has difficulties through theft of refined products. In 2004, 37 persons died when a truck illegally carrying 150 drums of gasoline exploded at a location a few miles from the Angolan capital Luanda.

Antiblaze®

Family of phosphorus-containing flame retardants manufactured by the Albermarle Corporation whose HQ is in the Netherlands. An example is Antiblaze® 182, which has found wide application to polyurethane foam products as well as to polyesters. A sister product is Antiblaze V6, also containing phophorous, which is used in soft furnishings and car upholstery, having been tested against applicable US and European standards for such applications. Others in this family of proprietary phosphorus-containing fire retardants include Antiblaze®125, widely used in carpet underlay, and Antiblaze® 195, also used in automobile upholstery. There is also an Antiblaze® product based on **ammonium polyphosphates**.

Anti-explosion pad

This uses the same principle as a simple mesh in retarding combustion rates by uptake of heat and radicals, but has an enhanced surface area by reason of a layered structure. It might typically contain two or more layers of mesh and metal foil: beads or even shavings of metal might also be incorporated. Configurations having a surface-to-volume ratio of up to 250 ft^{-1} ($\equiv$ square foot of metal area per cubic foot of pad) can be hoped for with a state-of-the-art anti-explosion pad. By way of perspective, this is approximately the surface-to-volume ratio of a single sphere of radius 4 mm. The layered structure therefore achieves an effect equivalent to that of reducing the metal to pieces of that size. A good porosity is also required and values of this in excess of 80% are common in such devices.

Antimony compounds, action of as fire retardants

These often consist of antimony oxide plus a halide of antimony. In the combustion of a solid such as a plastic, pyrolysate burns above the surface and is restricted in doing so by the step:

$$H + Br \rightarrow HBr$$

that is a halogen atom, released by the antimony chloride, removes a hydrogen atom which would otherwise have contributed to acceleration of combustion. The antimony *oxide* promotes char formation in the solid or, equivalently, inhibits breakdown of the solid into ignitable vapours. Other antimony compounds used as retardants include **sodium antimonate**.

Antistatic paint

In situations where liquid is being stored, the vapour/air mixture above its surface is an ignition hazards for which static electricity can provide the ignition source. This can be overcome by use of an antistatic paint (a.k.a. conductive paint) on the inside surface of the tank. Such a lining will ensure that any incipient electrical charge is conducted away so that spark formation is precluded. The term "static dissipative" is sometimes applied to antistatic paints. Antistatic paints are sometimes epoxy based, as with EPSP 393, manufactured by Osaka Paint, which is recommended for use in tanks of crude oil or of any petroleum distillate or residue. Antistatic paints are used not only in storage vessels but also for walls and ceilings where hydrocarbon vapour hazards exist. FPFAS 513, also from Osaka Paints (also epoxy based), is intended for such use. Where walls and floors are coated with antistatic paint, earthing is necessary and this might take place naturally through the building structure if there are steel supports, otherwise intentional connection to earth will be necessary as part of the protection process.

Aquabloc®

Fire-resistant product containing **gypsum**, manufactured by American Gypsum, the makers of **EagleRoc®**. Its surface is specially prepared for subsequent application of adhesive for attachment of tiles.

Aralite®

Protective fabric used in fire fighting apparel manufactured by Southern Mills in Georgia. It is composed of **Nomex®** and the stitching uses **aramid fibre**.

Aramid fibre

Polyamide developed by DuPont in the 1960s and a possible choice of jacket material for fire fighting hose, for example in Hotstop®, which is a Tyco product. Closely related products also finding application to fire hose are Kevlar® and **Nomex**, also developed by DuPont. Aramid fibre products also find application to protective apparel including the **fire fighter hood**.

ARFF

"Airport Rescue and Fire Fighting", a widely used and internationally recognised acronym. ARFF vehicles include the **Airwolf** and the **Carmichael Cobra**.

Argonite

Extinguishing agent composed of argon and nitrogen in equal proportions. It is 21% denser than nitrogen alone and 18% denser than air, so that once it is released into an enclosure it will have a significant residence time in the lower part of the enclosure. It has the major advantage of being totally non-aqueous. Hygood i2 is a Tyco product equivalent in composition to Argonite.

Arizona walnut

Botanical name *Juglands major*. Tree growing to a height of up to 40 feet and used as a **firewise plant** not only in Arizona but also in New Mexico. Arizona sycamore (*Platanus wrightii*) is also a **firewise plant** as is Arizona elder (*Alnus tenuifolia*), a.k.a. New Mexico elder.

Arsonists, traits of

Studies of the social and psychological profiles of those convicted of arson have identified a few factors common to many of them. One is lack of parental support and guidance during the formative years. Another is lack of a sense of well being, possibly linked to absence of formal education and of stimulation in employment. Alcohol abuse is common amongst arsonists. In particular, there are well documented examples of convicted arsonists initially showing promise after treatment who reverted to their former ways under the influence of alcohol. Additionally, arsonists are often spectators at accidental fires and in order to experience at least the initial thrill of such an occurrence might send a false alarm to the fire services or even seek part-time voluntary work with the fire services! There is also a clear gender bias: only about 12% of those convicted for arson in the UK are female. Offenders also tend to be young. (*See also* **Taegu**.)

Arson statistics, UK

In a typical week there are about 3500 intentionally lit fires in the UK resulting in 50 injuries and at least one death. The damage caused by arsonists in a year is worth £1.3 billion.

Asian forest fires, 1997

At a time which coincided with the collapse of the major Asian economies, forest fires originating in Indonesia spread to the extent of affecting directly an area of 7500 km² and a much larger area by smoke impact. As far away as Singapore and Kuala Lumpur, pollution from the fire necessitated hospitalisation of persons for breathing difficulties and eye irritation. Further effects included airport closures through visibility restriction and impairment of the growth of crops including coffee and palm trees cultivated for their oil. There was also of course loss of a huge amount of timber. A knock-on effect on the already precarious economy of Asia at that time was inevitable.

Ecological effects in affected areas including Sumatra were heavy and far reaching. For example, many species of butterfly are believed to have been totally annihilated and small animals upon which Sumatran tigers prey were heavily reduced in number with the result that tiger attacks on humans occurred. (There was a positive though as yet unconfirmed spin-off. The Javan tiger was declared extinct in 1970, but sightings of it have been reported since the fires, indicating that the obscure areas which had been the habitat of the survivors had been affected by the fires.)

There have been a number of suggestions of the cause, including the destruction of trees by fire to make room for the more lucrative palm oil plantations. A forest undergoing such a change is called a conversion forest and such conversion is only illegal if carried out without the necessary approval. In the Indonesian forest industry illegal operations abound and theft of timber, which ultimately finds its way to another country, is very common. We note in conclusion that the Indonesian *oil* industry also suffers heavy losses from theft.

Aspirated foam nozzle

This works by drawing atmospheric air into a fire fighting nozzle in order to expand the foam. One would not apply aspiration to a high expansion foam if the maximum achievable expansion was desired. Aspiration is more common for low- and medium-expansion foams and it is usual for an aspirator to be available as an accessory to be attached to an existing nozzle.

Astro Damper

Trade name for a particular **flame damper** manufactured in the UK. It works by **intumescence**, comprising a grid made of PVC and graphite which, on heating as a result of an accidental fire, swells sufficiently to close all spaces previously used for air passage. It thereby seals off the duct in which the grid is installed. The manufacturer makes the material available in strips which can be used to construct grids to required sizes.

ASTRO XT Clamp

Whereas a **cast-in-place firestop**, by definition, is installed in a new building, modifications to buildings sometimes require its equivalent to be retrofitted. Devices including the ASTRO XT Clamp have been developed for this. A few centimetres of cable ducting protrudes beyond a wall through which cables are to be passed. The ducting is coated with an intumescent substance and from there the *modus operandi* is just as for a **cast-in-place firestop**. Recalling that a **cast-in-place firestop** only has a depth equivalent to that of the concrete, rarely more than 10 cm, the visual effect of an above-ground counterpart such as the ASTRO XT Clamp is quite minimal.

Asuncion, Paraguay

The scene in 2004 of a supermarket fire, the death toll from which exceeded 450. Inadequacies both in fire protection at the supermarket in the first place and in the response of the emergency services are evident. Doors for escape in a fire had been locked. The task of the fire services was made difficult by poor equipment, including hose which leaked through its walls. Because there were too few ambulances available, privately owned trucks were used to take the injured to hospital. One of the supermarket owners was charged with involuntary manslaughter.

ATC

Alcohol-resistant aqueous film-forming foam (AR-AFFF), a Thundercraft™ product. It is suitable for gasoline fires and also for fires in oxygenated organics such as acetone. It expands by about a factor of four, or with aspiration up to about a factor of ten. It has been used in situations where **frothover** has been evident.

Atherstone on Stour

Town in the English midlands, the scene in November 2007 of a warehouse fire in which four fire fighters were killed. In terms of loss of lives of fire fighters, it was the worst incident in the UK for 35 years.

Atlanta, Great Fire of

This occurred on 21 May 21 1917. There was one fatality and property losses amounting to about US$5.5 million. The fire began late morning on a warm day with a southerly breeze. There had been a number of calls to the fire department over the previous hour or two to widely separated parts of the city, and the crew attending a warehouse containing burning mattresses could not deal effectively with it as most of the fire fighting equipment was elsewhere.

The fire originating at the warehouse caused wooden homes nearby to start burning and propagation from there was rapid. Fire fighters later used dynamite to demolish rows of wooden buildings thus creating a fire break (as had been done in fires resulting from the San Francisco earthquake 11 years earlier) and this was effective in preventing further advance of the fire. By this time it was dusk; almost 2000 buildings had been destroyed and 2 million gallons of water had been used by the fire fighters. Rebuilding was over a good number of years, there being empty and derelict parts of the city in the meantime, some of which were not built over but landscaped. The Martin Luther King Memorial now stands on one such site.

Also in Atlanta was the Winecoff Hotel fire in 1946 which, having claimed 119 lives, was one of the most severe hotel fires ever to have occurred in the US.

Atlantic Eagle

Anchor handling, tug and supply vessel built in Canada and registered to that country for service in the North Atlantic. It has FiFi I/II classification, there being two pumps and two **monitors** with a combined water release capability of 7200 m^3 per hour; foam is also carried. The "throw" from each **monitor** complies with the DNV requirement of 180 m.

Atlantic Empress

Oil tanker which in 1979 was in a collision with another such tanker the *Aegean Captain*. There was heavy loss of life through the fires which resulted.

Attack apparatus

Term applied to a fire truck from which water is directed at the fire, that is it is at the fire fighting interface. It can be supplied with water from a **water tender** possibly via a **dump tank** in what is known as "**dump-and-run**" mode. In aerial fire fighting the term "attack" is applied to an aircraft from which water is directed at a fire.

Augusta, ME

The Light of Life Christian Ministries Building in Augusta was the subject of some difficulty and an official follow-up when it was claimed that it did not conform to fire codes. If that is so, it was an **out-of-code** building from its very opening, yet it had received approval from the code-enforcing authority.

Automated External Defibrillator (AED)

"Ventricular fibrillation" is a state where the muscles of the heart malfunction. This can be prevented from developing into **cardiac arrest** by prompt application of an electric shock, a process called defibrillation. An AED can be carried by a fire fighting crew. These are able to assess the condition of a victim from the heart rhythms and to override an attempt to apply a shock where ventricular fibrillation is not in fact occurring. The London Fire Brigade made provision for the acquisition of a number of AEDs in its 2006 budget.

Automatic direct injection, foam proportioning by

Water for extinguishment passes through a flow meter which sends an electrical signal to a pump through which the foam concentrate passes. On the basis of the signal, the concentrate pump adjusts its output and a constant ratio of water to foam concentrate is achieved. Unlike an **eductor,** this type of foam proportioning device requires an electrical supply, in fact with a heavy current ($\approx$ 50 A). Inverted, however, this point becomes an advantage: because energy is supplied electrically it is not taken in significant amounts from the pressure energy of water exiting the hydrant, avoiding depletion of such pressure energy and therefore affording greater "throw" on exit.

Automatic distress signal unit (ADSU)

An ADSU, when incorporated into a fire fighter's personal apparatus, will activate an alarm when there has been no bodily movement on the part of the fire fighter for a preset time, typically 30 s. This is a "man down" situation. In its simplest form, the ADSU will simply emit an audible signal. One way in which an ADSU can work is for there to be a column of mercury which if horizontal will complete a circuit connecting the alarm to a battery. Other fire fighters hearing this will respond by attempting to locate the body and will be aided in this if the ADSU also causes a light to come on, enabling senses both of hearing and of sight to be deployed in finding the "distressed" member of their team. The time over which such a search occurs is called the "seek time". The **sound alert localiser (SAL)** has been developed with a view to overcoming this difficulty. Some such devices including the **Pak Tracker™** have, in addition to noise and visible light emission, radio wave emission at a particular frequency which can be monitored at a hand-held receiver of which the rescuer is in possession.

Autopremier

Vehicle-carrying vessel, built in 1997, upon which there was a fire in July 2001 caused by breakage of a fuel line. When the fire was discovered, the vessel was 35 miles from the Welsh coast on a journey from Ireland to Spain. It was towed to Milford Haven, but not before a fire crew had been taken out to the vessel by helicopter to confirm that the fire was extinguished.

Avalon, NJ

The scene in December 2003 of a fire involving condominiums, assigned the high value of 10 on the **alarm number** scale. Thirty-nine "condos" were destroyed. Two appliances belonging to the Avalon Volunteer Department were amongst those sent and each (one of them a **hydraulic ladder**) was damaged.

The American-Darling B-84-B fire hydrant (courtesy ACIPCO)

The Ultimate Bandanna™ (Courtesy Hot Shield USA)

B

Back Bay

This up-market part of Boston, MA was the scene in 2007 of ladder failure on a recently acquired fire truck. The steel ladder bent as fire fighters ascended it. The call had been to a suspected fire not an actual one, and by means of a crane the ladder occupants were rescued. The truck was one of four recently purchased by the Fire Department, and engineers from the manufacturer who went to Boston to investigate the incident examined the other three and approved their continuing in service.

Bahrain, fatal fire in 2007

There were 16 fatalities, all of them expatriate construction workers from India, when fire occurred at a residential site for such workers in Bahrain in November 2007. Eleven other occupants of the residential site received injuries. It is believed that the fire had an electrical origin. India herself registered a protest to the effect that its nationals had been endangered by overcrowding. The bodies were flown back to Chennai (formerly Madras). Earlier in the year three Indian expatriates in Kuwait died in an apartment fire also of electrical origin.

Bald cypress

Taxodium distichum, tree widely occurring in the US along the east side from New Jersey to Florida and inland to states including Indiana and Illinois. Some of the oldest examples known extend to 40 m in height. The species is classified as a **highly flammable tree**.

Balmoral Sea

Support vessel in use in the Gulf of Mexico destroyed by fire in June 2000 whilst stationary in the "Industrial Canal" – a 9 km waterway – in New Orleans. All occupants of the ship were evacuated to safety. The vessel was being prepared for repairs and it is believed that a welding operation might have started the fire. In spite of the involvement of many conventional fire appliances and the fire fighting vessel **General Kelley**, the *Balmoral Sea* could not be saved and about eight hours after the fire started was almost completely submerged in the water.

Baltic Eider

Cargo vessel registered to the Isle of Man. Whilst the vessel was sailing in Danish waters in August 2001, there was a fire in the engine room, the cause of which was leaked fuel oil and its contact with hot engine surfaces. Extinguishment of the fire was achieved but not before significant damage had been caused.

Bambi bucket®

Perhaps the best known **monsoon bucket**, designed and manufactured by SEI Industries in British Columbia. It is available in the capacity range 275 to 9800 litre and foam can be added to its contents. Traditionally a **monsoon bucket** is filled simply by dipping, which requires a minimum depth at the water source. SEI have made available a pump for attachment to a Bambi bucket® and this enables the bucket to be filled at a source as shallow as 1 foot, filling times being up to about 3 minutes. This device can be supplied with a new bucket or retrofitted to an existing one. Sometimes a **monsoon bucket** is filled not with adventitious water but from portable tanks set up on the ground, such a tank being the equivalent of a **dump tank** in ground fire fighting. One such is the Fireflex Pumpkin Tank™, specifically designed by SEI for refilling the Bambi bucket®. Like the **Fireflex Low Profile Tank™** the "Pumpkin" has a float collar device and is frameless. The very largest in the Bambi bucket® range can be completely filled from the Heliwell™ tank, also an SEI product, which does have a frame and needs assembling by two or three persons. It can hold 14 900 gallons, one-and-a-half fills for the 9800 litre size of Bambi bucket®.

Bandana (or Bandanna)

In fire protection, apparel which provides facial protection from heat, for example in fighting a brush fire. A secondary function is the filtering of particulate matter from the fire to prevent its inhalation. An example is the "Ultimate Bandanna", manufactured by Hot Shield,[5] whose HQ is in California. This is made of **Carbon X**.

Banshee™

Range of sounders suitable for fire protection application. It is available in high-frequency (2.6 kHz) or low-frequency (900 Hz) forms and has volume

---

5   This company is owned by fire fighters.

control across a 15 dB range. It is "intrinsically safe" in that it draws current of the order of milliamps and therefore cannot provide an ignition source for a flammable gas or vapour.

Barbados II

One of two tugs owned and operated by Barbados Port Inc. Its duties expressly include fire fighting, for which it is equipped at **FiFi 0**, level having a single **monitor** with a capacity of 1200 m³ per hour.

Barnum and Bailey circus, fire at

At this circus in Hartford, Connecticut in July 1944 the circus tent caught fire. The death toll was 163, about half of them minors. It was just after the Second World War and allocation of fireproofing materials such as **borax** was made by the army, to whom an unsuccessful application for a quantity had been made by the circus operators. The tent had therefore not been fireproofed. Not all of the dead were ever identified.

Barracuda

ARFF vehicle built by Reynolds Boughton in the UK. It is available in two-axle (the Barracuda 4) or three-axle (Barracuda 6) form. Engines for either are up to 700 hp and up to 1600 gallons of water can be carried. The Barracuda 4 is in use, for example, at the airport at Abu Dhabi. Users of the Barracuda 6 include Birmingham Airport, UK where six are in service. Other provincial UK airports including Cardiff and Bristol use the Barracuda 6. Reynolds Boughton also manufacture an **ARFF** called the Marlin, in use at places including Hong Kong Airport. This has three axles and an engine of up to 1005 hp.

Barricade® Gel

Fire-blocking gel also usable in extinguishment, similar to **AFG Firewall™** and to **Thermo-Gel®**. The polymeric constituent is potassium polyacrylate. As with other such products, it should be removed by vigorous hosing after the fire danger is over.

Basotect®

Product of BASF, foam made from **melamine**. It has been chosen for "acoustic insulation" in the new Boeing 787 aircraft and requires no added retardant.

BC30

Dry extinguishing agent containing 98% of **sodium bicarbonate** at particle sizes < 350 μm. In addition to the primary extinguishing effect, which is carbon dioxide release, BC30 and other bicarbonate preparations remove free radicals in the combustion process by absorption on to the particles before they decompose. This is analogous to "wall" removal of radicals in studies of gas-phase combustion processes.

Beijing, recent fire in

An inauspicious event occurred in July 2007 when a fire broke out at the Beijing Olympic facility a full year before the Games were due to commence. The fire was in a partly constructed gymnasium and was attended by 17 fire appliances. There were no deaths or injuries. A news release tentatively attributed the fire to "improper construction practices". A reader should note that fire protection for the Games facilities in Beijing has involved organisations from outside China including Ove Arup.

Below horizontal operation of aerial ladders

Imagine that there is a fire accessible only by a bridge at a higher level than the site of the fire, necessitating access from the bridge itself for rescue. An aerial ladder so used is then said to be in below horizontal operation. For example, **Metz** manufacture an aerial ladder which can be used at up to 15° below the horizontal with the appliance itself horizontal . The entire appliance can however be tilted a further 7°, giving a total angle of 22°.

Benoni High School, Johannesburg

The scene in 2007 of a fire during school hours. It began in the store of the school's woodwork room which, of course, would be the part of the school with the highest concentration of combustible material (the highest "fire load"). Two woodwork teachers who themselves tackled the fire subsequently received hospital treatment for smoke inhalation.

Bergenfield, NJ

The scene in December 2005 of a gas explosion in an apartment block in which three persons died. Excavation to remove an oil tank close to the apartments was taking place and severance of a gas pipe during that is believed to have been the origin of the accident. Two of the dead were in their seventies. Some controversy arose from the fact that a fire fighter who was aware of the gas leak before the explosion decided not to evacuate the building.

Beriev Be-200

Aircraft which can at manufacturing stage be configured as an **air tanker**:

most air tankers are made by adaptation of an aircraft previously in passenger or freight service. It is manufactured by the Berieva Aviatsionnyi Kompaniya (Beriev Aviation Company) in the Ukraine and made its western debut at the 1999 Paris Air Show. The Be-200 **air tanker** is powered by two jet engines and is amphibious. Its payload is a little over 3000 gallons of water, making it a competitor of the **P3 Orion** and the **C-130 Hercules**. As yet its only use outside the former Soviet Union has been only in two places: Greece, where one loaned from Russia was used to fight the **Peloponnese** forest fires, and Indonesia, for use in Sumatra and Kalimantan. Indonesia is however expected to buy two Be-200s.

Berkan B

Cargo vessel detained in the UK port of Ipswich in 2005, having been deemed unfit for service on safety grounds. The vessel was registered to Georgia. The Maritime and Coastguard Agency have the legal power to detain a foreign ship in such circumstances. One concern was that each **fire damper** in the engine room was seized and therefore not capable of operating. Others were lack of knowledge of fire procedures amongst the crew and blockage of escape exits. After 11 days of detention the vessel was allowed to depart on a single voyage to Istanbul for repairs.

Bethlehem, PA

The scene in February 2007 of the death of a fire fighter from **cardiac arrest**. The deceased had in the afternoon attended a house fire, and started to experience symptoms in the evening whilst still on duty but back at base. He was taken to hospital and died there.

Beverly Hills Supper Club fire

The destruction by fire of this venue in Kentucky in May 1977 is on record as the worst nightclub fire in US history. There were 165 deaths and 200 non-fatal injuries. There were approximately 3000 persons inside the club, distributed across a number of rooms, when the fire started. Waitresses entering one of the rooms at about 9:00 pm observed smoke, and the fire department were sent for and arrived very promptly. There were no fire alarms, nor was there a sprinkler system, and advance of the fire was very rapid. In one room a comedy act was taking place. When it was interrupted to warn patrons to evacuate because of fire, this was seen by some as part of the act and consequently dismissed! When very soon afterwards it became obvious that the warning was in no way in jest, there was total disorder as persons attempted to escape without supervision from the club management or the fire department, and such exits as there were became blocked. In the investigation it was revealed that the building had over the previous few years received a number of extensions without any regard to fire safety and that an unlicensed architect had overseen some of them. The club was never rebuilt.

Biddulphsberg

Scene of a battle during the Boer War, the outcome of which was eventually influenced more strongly by a veldt fire than by the actions of either side in the conflict. British soldiers were receiving gunfire from the Boers and were initially in some degree protected by the high grass which made for concealment. A grass fire from the west propagated in their direction and was accelerated by wind. The dependence of the burning speed of such a fire on the wind speed is much stronger than linear, perhaps exponential, and many of those wounded by enemy gunfire were not able to make it to safety and were fatally burned. The Boers had been protected from the grass fire by a road which had acted as a fire break. British deaths due to the veldt fire numbered 38; the bodies were buried *in situ*.

Biodiesel, electrostatic hazards with

On this front it appears that the news is all good! Most biodiesels have better conductivities than mineral diesel and do not require a **fuel conductivity improver**. For any which might, those which have been developed for mineral diesel are effective. Where biodiesel and mineral diesel are blended, the biodiesel itself acts as a fuel conductivity improver.

Biodiesel production, fire during in 2006

Biodiesels are becoming more and more widely used, largely because of their carbon neutrality. There are very many examples of accidental fires in the production of conventional liquid fuels and one expects intuitively that as biodiesels proliferate they will not do so without mishap. At a biodiesel plant in Bakersfield, CA in 2006 there was a fire involving methanol. This is widely used as a reagent in biodiesel preparation, as esterification with methanol is a means of raising the cetane number of a biodiesel. The cause of ignition is believed to have been static electricity. There was total destruction of the building. Much of the chemical inventory – methanol and unprocessed biodiesel – was in railcars at the plant at the time the fire began so it was possible to move it to safety. There were no deaths or injuries. In the same year there was a fire involving stored biodiesel in **Salem, OR**.

BioSOY™

Hydraulic fluid made from vegetable oil, having a flash point of 208°C and a viscosity at 40°C of 34.2 centistokes. On the basis of these quantities (although there are others to be considered, including the variation of viscosity with temperature) we might consider it comparable to **Nuto H32**.

Bizon™

Flame-retardant material for incorporation into plastics and artificial rubbers, most suitable when moulding temperatures do not exceed about 220°C. Like antimony compounds and **borax** (from each of which it is in chemical composition totally different) it acts by forming a char layer at the surface.

Black walnut

Botanical name *Juglans nigra*. This tree, the timber from which is prized, is classified as a **highly flammable tree**. It occurs along eastern North America from Ontario down to Florida and in parts of south western USA, including central Texas.

Blazemaster®

Gloves for fire fighter use of UK manufacture. The outside is made of leather and the inside of Kevlar® and **Nomex®** in a fabric having the structure of towelling.

Blossburg, PA

The scene in 1967 of an air crash originating with an on-board fire. The aircraft, a BAC 1-11, was on its way from Elmira, NY to Washington DC. Air intended for the aircraft's auxiliary power unit was misdirected through valve failure and came into contact with fuel vapour. The fire so initiated spread to the aircraft's hydraulics. The aircraft crashed with the loss of 34 lives.

Bluejoint reedgrass

Just as some plants are known to provide limited protection from fire and are therefore called **firewise plants**, some others have been identified as such strong burners that their growth close to property is advised against. Bluejoint reedgrass (*Calamagrostosis canadensis*), which occurs in parts of North America including Alaska, is a case in point. It is a tall grass and especially susceptible to fire during the spring season.

BOAC[6] fire, 1968

A BOAC Boeing 707 aircraft taking off from Heathrow Airport in London on 9 April 1968 experienced detachment of one engine, which having caught fire dropped over the part of London close to the airport. The aeroplane was going to Sydney, with its first stop at Zurich. The aircraft was returned to the ground. By the time it was stationary there was also fire in part of one wing and evidence of melting of parts of the structure close to it. Fire spread to other parts of the aircraft and escape chutes were damaged. Help from the airport's own fire appliances was delayed. The five fatalities included one female member of the cabin crew. The accident was attributed to failure of a rotor in the engine which initially caught fire.

Boeing 737, two incidents with

In 2001 an explosion occurred on a Boeing 737 whilst it was on the ground at Bangkok Airport, and similarity to an incident with a Boeing 737 at Manila Airport in 1990 was noted. The two are believed to have the following common cause. In each case the aircraft was on the ground with air conditioning in operation and the centre fuel tank, positioned under the fuselage, empty. Heat removed from the cabin by the air conditioning is taken to a heat exchanging surface close to the centre fuel tank for radiative and convective loss from there. With a tank only partly full the quantities of heat so transferred would be insufficient to affect in any measurable way the tank contents with their huge thermal mass. In a so-called empty tank, however, there is residual vapour which *can* be affected by amounts of heat of this order, especially if the pump, which is immersed within the fuel, is turned on. It is in fact believed that the pump can provide an ignition source if it is operating under dry conditions as it is designed to be lubricated by the liquid fuel which passes through it. Boeing themselves had always recommended that this pump be turned off when the centre fuel tank was empty, and a reminder was circulated after the Bangkok accident (in which there was one fatality). The 737 which caught fire at Manila Airport is known to have been wired with **Kapton®** and in at least one "blog" on the topic a connection has been made.

Boeing 747 Evergreen Supertanker

The largest **air tanker** in existence, made from a Boeing 747-200 previously in airline service. It has about twice the water holding capacity of the **Tanker 910** and eight times the water holding capacity of the widely used **P3 Orion**. A difficulty with very large aircraft adapted to fire fighting

---

6   British Overseas Airways Corporation, later subsumed into British Airways.

is that the heights required for them to fly safely cause water, if released under gravity alone, to disperse too widely, most of it missing the target fire. With the Boeing 747 Evergreen Supertanker this problem has been addressed in two ways. First, water release does not depend on gravity alone but is pressurised, enabling the supply to be focused. Secondly, modification to the wings was effected whereby the aircraft can fly at heights well below 1000 feet. There are economic as well as operational benefits from the use of one large air tanker in place of half a dozen or more smaller ones.

Boeing 747-400, tail fuel tank of

In contrast to the earlier series 747s, the 400 series was built with an additional fuel tank at the rear. In 1998 a directive to airlines flying the 400 series was issued whereby the tail fuel tank was not to be used. Also, none of the tanks was to be allowed to run dry in flight. This had the effect of reducing the range of the aircraft, or of permitting an unchanged range with less weight and therefore fewer passengers.

Boeing 777 crash, Heathrow, January 2008

The was no fire or explosion as a result of this accident, but points relevant to fuel hazards did feature in follow-up and one important lesson was learnt. Fuel did in fact leak from the crashed plane, but without igniting. Spar valves, closure of which stops fuel supply to the engines in the event of an emergency, were fitted to the aircraft. When the crashed 777 was examined, both of these were found to be in the open position and this was the reason for the fuel leakage which, in the event, had no consequences, but which if the spar valves had fulfilled the purpose for which they were installed would not have occurred. The Aircraft Accident Investigation Board's Bulletin made the following points. Electrical power for spar valve closure is provided when the fire handle, which initiates release of extinguishant (held in "fire bottles") in the engine areas, is engaged. Such power is also provided when the fuel control switches are placed in the "cut-off" position. With the 777 as manufactured, when both fuel control switch and fire handle are operated, it needs to be in that order – fuel control switch first and fire handle second – for the highest reliability of the spar valve closure. For a reason clearly identifiable when the circuitry is reviewed, this reliability is lower if the the two operations are reversed. Accordingly, Boeing had issued a service bulletin recommending a simple circuit modification such that the dependence of the reliability of spar valve closure on sequential operation was eliminated, and the FAA requires that to be implemented in all 777s in service in the US by July 2010. It had not been implemented in the 777 which crashed at Heathrow, the evacuation checklist for which recommended that the captain should operate the fuel control switches and the first officer the fire handle. Such independent operation tends to preclude sequencing. Boeing have issued a safety recommendation that the need for sequential operation be noted on evacuation checklists in 777s not yet having implemented the modification recommended in the service bulletin issed before the Heathrow accident.

Boilover

Phenomenon sometimes observed in fires involving stored liquid fuels of wide boiling range. It is the result of instability within the liquid brought about by contaminant water which, being denser than the fuel, will be at the base of a storage vessel. Nucleate boiling (as opposed to film boiling) of water occurs when the temperature is above the boiling point of water but below that of the organic phase with which it is in contact, which therefore acts as a heater surface. A stage is reached where all such water evaporates and becomes superheated, that is no longer in phase equilibrium with liquid. When such superheated vapour exits the liquid, the effect is similar to that of placing a stirrer in the liquid, enhancing burning rate and possible loss of containment so that the liquid fire is extended to sites outside the vessel. Boilover has occurred in fires involving crude oil, as at **Milford Haven** and at the **Showa Oil Niigata Refinery**. A milder effect than boilover having the same cause is spillage of liquid over the tank walls to the outside, known as **frothover**. This was observed, for example, at the crude oil fire in Newport, OH. It is possible for an effect akin to boilover to occur if extinguishment water enters a burning liquid. This is known as **slopover**. (See also **Toa Oil**.)

Bonassola

Anchor handling, tug and supply vessel classified FiFi I/II, an example of a vessel having dual classification. It is currently operating in Italian waters. Italy's offshore oil and gas activity is mainly in the Adriatic.

Booster fire hose

Hose capable of withstanding pressures involved in the drawing of water from a **booster tank**. It is commonly a three-layer structure, being synthetic rubber on the inside covered with a woven material over which is a smooth synthetic rubber outer coating. Such hose is pressure tested to not less than 400 psi: hose for low-pressure use such as **Vinylflow** will have a containment limit of about 70 psi.

Booster tank

Receptacle by means of which water is carried by a fire truck. The name derives from the idea that water in the tank will be used to supplement water supply at the scene of the fire, that is such water is a secondary supply. However, frequently it is used over the short time that it takes for the water supply at the fire to come on line. In some fire department responses, e.g. where a car is on fire, the contents of the booster tank might be sufficient for extinguishment, in which case it is of course the primary supply. Being delivered at relatively high pressures, it requires stronger hose than water being taken from a **water tender** or a **dump tank** and this is known as **booster fire hose**.

"Borate bomber"

In modern aerial fire fighting the water dropped form a helicopter or a fixed-wing aircraft contains a suitable foaming agent, often **AFFF**. In the earliest days of the use of aircraft for fire fighting borax was added and this led to the term "borate bomber" as a synonym for "water bomber", a.k.a. **air tanker**. **Borax** will enhance the extinguishment capability, but in fact the primary reason for its inclusion was to elevate the boiling point to reduce water losses by evaporation as the released water approached the fire.

Borax

Sodium borate, $Na_2B_4O_7.10H_2O$, a mineral the uses of which include fire retardancy, for example in wood. Its action is believed to be in the promotion of char formation or, conversely, the inhibition of volatile release. The effectiveness depends on the pH, which is why borax is sometimes combined with boric acid as the presence of the latter enables the pH to be controlled. Boric acid alone is sometimes used as a fire retardant. Boron preparations are available which are suitable for incorporation into polymers at the manufacturing stage to increase their resistance to fire. Products marketed as Firebrake are examples and contain zinc salts of boron; such a salt will also act as a **smoke suppressant** and in some products, including **ZB-223**, this is its primary role.

Bournemouth, UK

This south coast resort town was the scene in July 2002 of an accident involving a **Sikorsky S61** helicopter. After fire had begun in one engine during flight, the crew landed at Bournemouth Airport and evacuated. The time between the initial awareness of the fire by reason of the instrumentational warning and the evacuation of the crew to safety was a mere 82 seconds. The helicopter itself was totally destroyed in the resulting fire.

Bow Mariner

Vessel carrying a cargo of ethanol, which exploded off the Virginia coast on 28 February 2004, with 21 deaths amongst the crew. The vessel had also been carrying methyl tertiary butyl ether (MTBE) which, prior to the accident, had been dropped off at Linden, NJ. The vessel was 26 years old at the time of the accident and had left Saudi Arabia with her payload of ethanol and MTBE. Having dropped off the latter at Linden, she was proceeding to Texas City to deliver the ethanol. At the time of the accident therefore she was carrying ethanol as well as both light and heavy fuel oil for her own use. It is believed that between the departure from Linden and the accident, the tanks previously containing MTBE had been opened for cleaning. It was noted in the follow-up that the tanks once emptied of MTBE had not been "inerted". The report to the International Maritime Organisation (IMO) from the Virginia Coastguard attributed the explosion to ethanol vapour in air.

Boyd, James

Successful applicant for a US patent, granted in 1821, for fire hose lined with rubber having a woven cotton outer structure. At that time fire fighting with buckets passed along a line of men (a "bucket brigade") from the source of the water to its point of application to the fire was still taking place in some parts of the US. The last man in such a line had the extra duty of throwing the water, hopefully effectively, on to the fire and sending the empty bucket on its return for refilling. The inefficiency of such a *modus operandi* had by Boyd's time been quantified: it was contended that 100 feet of hose was equivalent to 60 men with buckets. Incentive for the development of strong and reliable hose was therefore considerable.

BP® Hydraulic AW

Series of hydraulic oils, mineral oil based, with viscosities (40°C) in the range 32.5 to 63.2 centistokes and flash points in the range 199 to 249°C. They are therefore of higher viscosity than the **Nuto** range, the most viscous of which has a value of 32 centistokes at 40°C, but the flash points of the BP® Hydraulic AW and Nuto ranges are broadly the same.

Bradford football stadium fire, 1985

Fifty-six lives were claimed at this fire, which is believed to have begun when a discarded cigarette caused ignition of litter which had accumulated over many years. The part of the stadium in which the fire began was made of wood and had first been used in 1908. The fire was enclosed on three sides, that is in all directions other than that of the pitch where many spectators did escape to. There was flashover within the enclosed area and subsequent fire spread was exacerbated by the presence of a bitumen coating on the roof.

Breathe-tex®

Material used as a **moisture barrier** in fire fighting apparel. The International Association of Fire Fighters (IAFF) have expressed concern at failure of this product due to breakage[7] and a recall was initiated.

Breathing apparatus, failure of in a fire in 2002

What began as a fairly routine call-out to a commercial building in St. Louis, MO in May 2002 in fact resulted in the deaths of two fire fighters, and in the legal follow-up the tragedy was attributed to failure in an **SBCA** and the manufacturers of it were required to pay heavy damages. One of the fire fighters died after removing his breathing apparatus in an attempt to free a valve within it. This fire fighter was at that stage helping another one who had become incapacitated and whose **PASS**, from the same manufacturer as the **SCBA**, was not functioning. Punitive damages as well as actual damages were required from the manufacturer.

Brigadier fire hydrant

Like its **American Darling** and **Waterous** equivalents, this fire hydrant, which is manufactured by Clow Canada, has a 5.25 inch valve opening. The original company, before acquisition by Clow, was called McAvity and hydrants bearing that name are still common in Canada, some of them having seen close to a century of service. Fire hydrants are very much classical devices and modifications over time of the order of decades have been minor. They include the use of new materials including polyurethane for certain components. Possibly the oldest fire hydrant in the world is that at **Nara**.

British Library book storage facility

This building in Yorkshire, which stores seven million volumes and is expected to expand, is protected by **OxyReduct®**. Another important building in which OxyReduct® has been installed is the British Airways IT Centre at Heathrow Airport.

Brominated polystyrene

A widely used flame retardant for polymers, resins and fibres. Being poly-

---

7   http://www.iaff.org/HS/Alerts/alert06.asp

meric itself, it has no mobility within the matrix of the substance to which it is added. Polystyrene can of course be brominated to various extents, and proprietary forms of differing composition are available. A typical composition is 50% bromine, meaning that the effect of the bromination has been:

$$C_8H_8 \rightarrow C_8H_{6.67}Br_{1.33}$$

Bronx, apartment fire 2006

The FDNY responded to this fire on 1 February 2006 by giving it the high classification of 6 on the **alarm number** scale. Fire fighters' efforts were made more difficult by strong winds. Several residents was hospitalised and 21 fire fighters treated for smoke inhalation.

Bronze, use of in fire fighting plant

This alloy is the preferred material for fittings in fire fighting plant where seawater rather than fresh water is used, for the obvious reason that it is relatively resistant to corrosion. Notable amongst its applications is fire protection at offshore oil and gas installations.

Brunsbuttel

This German town was the scene recently of a fire at a power station involving transformer oil. The ignition source is believed to have been electrical. At almost exactly the same time there was a transformer oil fire at a power station in Geesthacht in northern Germany.

Brush fire, intentional lighting of

A fire involving grass and other forms of vegetation can be initiated by accidental contact with a heat source, as is believed to have happened in Australia in 1983 when two copper conductors, part of the electricity transmission system, contacted and shorted, causing molten copper to drop on to the (largely dead) vegetation below. If the cause is accidental, the fire will have begun at a single site, so if a brush fire begins simultaneously in several places distant from each other, that is evidence that it was begun deliberately. This is believed to have happened in August 2007 at the Sicilian resort of Cefalu, where a brush fire began at about the same time at at least five widely separated places. There were no deaths or injuries, although there was significant property damage and a number of hotels had to be evacuated. The fire was accelerated by wind from the south. The fires at **Peloponnese**, which claimed many lives, are also believed to have been started intentionally.

Brushmaster

Short-wheelbase fire appliance from American LaFrance. Being intended for **brush fire** use, it has a pump and water but no ladders. It has a Darley pump and a water tank capable of holding 1000 gallons. Two new Brushmasters were recently delivered to the Fire Department in Loredo, TX.

Bryce Canyon, UT

The scene in 1947 of an accident involving a DC-6 aircraft and originating with a fire on board. The aircraft was on a flight from Los Angeles to Chicago and was attempting to land at the airport at Bryce Canyon after fire on board was discovered. The DC-6 was equipped to transfer fuel between tanks during flight in order to make the weight distribution even, and in such an operation vapour had leaked and had been ignited by heating elements in the cabin's own heating system. The fuel being used would of course have been in the gasoline boiling range, as the DC-6 used piston engines. The aeroplane crashed and all 52 persons on board were killed.

Buckinghamshire and Milton Keynes Fire Authority, use of a quint by

The **quint** is much less common in the UK than in the US. The Buckinghamshire and Milton Keynes Fire Authority do operate one with a Volvo cab/chassis and a ladder assembly made by Magirus (part of Iveco). The pump is one from the **Godiva World Series**. (*See also* **AllRounder**.)

"Buddy breathing"

Procedure whereby the air supply in a **SCBA** can be extended to a second user whose own air supply has become depleted. This requires the capability to split the air supply from the cylinder. Some **SBCA**, including the **North 800 Series™** have this capability as standard.

Buildings, fire hose for

Hose installed in a building and stored on reels or racks for use by the building occupants will be a less robust product than hoses carried by fire trucks for use by trained fire fighters, several examples of which have entries in this volume. Pressure within building hoses will be low, partly because of limitations imposed by the water supply, but also because hose operation at a pressure even as relatively low as 100 psi can lead to fire fighter fatigue, to which an untrained operator ought not to be subjected. A further point is that water application which is too copious can lead to water damage which possibly exceeds the fire damage. Some hoses are suitable for such use and, up to certain degrees of fire fighting challenge, fire truck use; **Sirocco 9303** is an example. By contrast **Rack Hose™** is for interior use only.

Bukit Ho Swee

Location of a squatter camp within Singapore, the scene in 1961 of a fire which claimed four lives and caused many (>80) injuries. The fire is seen as having a place in social history in that it showed that not only disease and infection but also extreme fire hazards result when human beings have to live under such conditions.

Bulldog

Tug with an additional fire fighting capability at **FiFi I** level. It is used to escort vessels bearing liquefied natural gas (LNG) on their approach to the terminal at Savannah, Georgia. The vessel was built by Washburn & Doughty Associates of Maine, who have previously supplied several vessels for such use.

Burgan oil field, accidental fire at

The Burgan oil field in Kuwait is the second largest oil field in the world. In May 2007 there was an accidental fire there involving one non-fatal injury. The only disruption was of LPG supply to a facility operated by the Kuwait National Petroleum Company.

Burn Barrier™ 10-10

Fire-retardant paint displaying **intumescence**. Solvent (not water) based, it can be applied to wood, masonry or metal surfaces after suitable preparation. Burn Barrier™ signifies a large range of fire-retardant products. Another is Burn Barrier™, which is used to fire retard fabrics both natural and synthetic. Burn Barrier™ X is used to protect Christmas trees, being formally approved in California for such use.

Burn patterns, importance of in fire follow-up

An irregular burn pattern – one where the debris from the fire is more severely affected by the fire than others close to it – can signify either arson using an accelerant or use of an incendiary device. Sometimes fires due to the former have initially been incorrectly attributed to the latter. This is not surprising as the two work along very much the same principles.

C

C-130 Hercules

Like its stablemate the Orion, the C-130 has found application as an **air tanker**. It has the same classification as the Orion, being capable of carrying about 3000 gallons of water. In June 2002 a C-130 **air tanker**, having dropped water on to a fire, then crashed over the town of Yosemite in northern California, killing all three crewmembers. This and a number of other such accidents in the early 2000s were one incentive for the manufacture of larger air tankers such as the **Boeing 747 Evergreen Supertanker** and the **Tanker 910**, which have respectively eight and four times the water holding capacity of a C-130 Hercules or a **P3 Orion**.

Cabinets for flammable liquids, principles of construction of

Steel is the usual choice of material for cabinets holding containers of flammable liquids, and the thickness ("gauge") is more important than the grade of steel, although some manufacturers of such cabinets do specify this, e.g. 304 stainless. Typically, 18 gauge (0.0478 inch ≡1.2 mm) steel is used for the double walls, there being a space of an inch or more in between. All seams should be welded. The cabinet can be connected to earth ("grounded") to eliminate electrical ignition sources. Three-point latching is common for cabinets with double doors. Shelves within the cabinet need to be strong, and galvanised steel is a common choice of material. Configuration of shelving will depend upon whether the liquid is being stored in a small number of large containers or a larger number of small ones. Where access is required frequently and the continual opening and closing of doors is time wasting, a **fusible link** can be installed. Broadly speaking, such cabinets from stock will have a liquid holding capacity in the 20 to 50 gallon range if standing on a floor, and in the 15 to 20 gallon range if wall mounted. Stackable ("piggyback") cabinets are also available. A **drum safety cabinet** will hold a single drum of flammable liquid or possibly (if the orientation is vertical) a pair of drums. For all such cabinets the total cabinet volume will be of the order of five times the volume of liquid which it is designed to store.

California Department of Forestry (CDF), loan of aircraft to

The CDF have obtained aircraft under the **Federal Excess Property Program**, having converted them for fire fighting. Although the Program remains in force, aircraft are not as plentiful under the Program as they once were and the CDF is budgeting for eventual replacement of these aircraft from other sources.

California, Texas and New York, Insurance Services Office classifications for

We would expect these three states to be amongst those with the best **Public Protection Classification (PPC™)**. Note that the figures for New York do not include New York City, which is unclassified.

State	Highest classification	Modal classification	Number of class 10 communities
CA	1 (12 communities)	5 (237 communities)	17
TX	1 (eight communities)	9 (483 communities)	184
NY	2 (eight communities)	9 (681 communities)	23

(See also **Rocky Mountain states, Insurance Services Office classifications for**; **Selected US states, Insurance Services Office classifications for**.)

Canadair 415

A.k.a. the Bombardier 415, an amphibious aircraft for fire fighting, capable of holding 1600 gallons of water. Its ability to land on water enables it to take on water from a lake as with the **Fire Boss**. It can however "scoop" water without itself landing on the water surface, meaning that water can be taken from rivers which are too narrow or have too many bends for landing. In addition to Canada and the US, France, Greece, Italy, Croatia and Spain have the Bombardier 415 in service. It has also recently been used in Corsica.

Candles, house fires due to

Once a means of providing lighting for homes, candles are now present in many homes for their decorative appeal; sale of candles rises steeply year by year. A candle releases heat at a rate of a few tens of watts and burns with a diffusion flame. In the UK in the year 2000 there were 10 deaths due to house fires having originated with a candle and 900 non-fatal injuries.

Car batteries, behaviour of in fires

A car battery can explode suddenly during a vehicle fire, causing release of chemicals and further hazards to victims and rescue workers. This was an issue at the vehicle fire at **Toxteth**.

Carbon dioxide extinguisher

Carbon dioxide for fire extinguishment can be obtained from **sodium bicarbonate** or **potassium bicarbonate**, although some contain gaseous carbon dioxide. Commonly, carbon dioxide in such an extinguisher is at a pressure of 100 bar or more in a steel or aluminium receptacle. The equilibrium vapour pressure of carbon dioxide at room temperature is about 60 bar. The carbon dioxide in the extinguisher is therefore superheated. However, the pressure drop on release can create conditions such that solid ("dry ice") does transiently form and the release is visible. A typical carbon dioxide extinguisher will contain 2 to 5 kg of the gas releasable through a funnel-like device. Carbon dioxide is also a component of **Inergen**.

Carbon fibre

These materials find application to fire engineering as one ingredient of certain composite materials used to make lightweight gas cylinders for **SCBA**. Other materials present in the composite are likely to include aluminium, as a liner, and a suitable resin. **Aramid fibre** products might also be incorporated. (*See also* **AirBoss**.)

Carbon monoxide, biomimetic sensors for

A biomimetic device is one which mimics biological processes. Carbon dioxide can be sensed by means of a synthetic chemical substance which behaves like haemoglobin in the blood. The substance changes colour when carbon monoxide reacts with it. If such a sensor is incorporated into a detector, this colour change is the basis of a response. Use of a semiconductor for carbon monoxide detection is also possible, as with the **TGS 813**. (*See also* **Electrochemical detection of carbon monoxide; Carbon monoxide, infrared detection of.**)

Carbon monoxide, infrared detection of

Carbon monoxide absorbs infrared at a wavelength of about 47 μm, in conformity with the Beer–Lambert Law. This is the basis of detector instruments for carbon monoxide, which as well as being poisonous is an indicator of incipient fire. Methane can also be detected by infrared as with the **IRcel®CH$_4$**.

Carbon X®

Fire-resistant material containing an **aramid fibre** as well as another fibre having an acrylonitrile structure. In the fire-resistant materials market it competes with products including **Nomex®**, Kevlar® and some of those having had the **Proban®** treatment. It was developed and is manufactured by Chapman Thermal Products in Salt Lake City, UT.

Cardiac arrest

This is the most common cause of deaths of fire fighters during duty. Some of the many recorded examples will be covered as case studies in this volume (including that at **Bethlehem, PA**, in which cardiac arrest occurred a number of hours after the victim had completed duty at the scene of a fire and returned to base). One contributing factor to cardiac arrest in fire fighters is high levels of carbon monoxide inhaled. Recently the need for any pre-existing heart condition in a fire fighter to be known and documented has been emphasised. This follows release of statistics by the NFPA, which reveal that over the significant period of time covered by the statistics three-quarters of the fire fighters who so died were themselves aware of a heart complaint.

Cardon Refinery

This refinery in eastern Venezuela,[8] owned by Petroleos de Venezuela S.A., was the scene recently of an explosion in which 16 persons were injured. The refinery has a capacity of 305 000 barrels per day and is in close proximity to other refineries in what is known as the Paraguana refining complex. The fire is believed to have been caused by contact of a spark with gasoline vapour.

Carmichael Cobra

Range of **ARFF** vehicles built in the UK, the top-of-the-range having a water pumping rate of 4000 gpm. Constructed on a stainless steel frame, the Cobra has panels made of **GRP**. A few days before this entry was written, a new Cobra was delivered to Edinburgh Airport.

---

8 Venezuela was one of the founding members, over 45 years ago, of OPEC and has been a member continuously throughout that time. By contrast the only other country from the Americas to be in OPEC – Ecaudor – was in late 2007 readmitted having once been a member and then withdrawn.

Carry-Lite

Fire hose manufactured by Mercedes Textiles in Canada, distinguished from many others in being "adhesiveless". The woven jacket and the liner are attached ("welded") by a patented process. In addition to operational advantages, this makes for increased life expectancy. It is designed for moderate (up to 300 psi) working pressures.

Cartersville, GA

The scene in 2005 of a fire at a brewery caused by leakage of ammonia, which was being used as a refrigerant. With an ammonia fire there is the additional factor of the toxicity of unburnt fuel necessitating HAZMAT procedures.

Cascade system

Means of filling a cylinder with gas often used in fire fighting to replenish the air cylinders of **SCBA**. The procedure involves series connection of a number of cylinders, a high-pressure supply cylinder being the furthest upstream. The supply cylinder will, in the US, conform either to ASME or DOT design standards and for a satisfactory fill to be achieved will need to contain air at high pressure, up to 6000 psi. The use of lower pressures can lead to incomplete filling of the vessel undergoing replenishment, which in a fire fighting situation can of course endanger the life of the user of a **SCBA** into which the cylinder is subsequently placed.

Caspian 2026

A hose of **textile-reinforced rubber** type, though differing from, for example, **Storex Laylite** in that it has two layers of synthetic rubber only the outer one of which is reinforced. Caspian 2006 has the ability to work with water at very high pressures, up to 580 psi, and is therefore suitable for use as **booster fire hose,** though its actual use is by no means solely as that.

Cast-in-place firestop

Such a firestop in a floor is installed before the concrete is poured. It has a supporting shell which when removed after the concrete has set leaves a layer of intumescent material, providing a safe entry/exit for plastic pipes. The intumescent material will on heating swell and cause the plastic pipe to cave in as with a **fire collar.** Examples of cast-in-place firestops include the **Fyre-Can** and **CP-680** ranges. Firestops are available which are suitable for use with metal pipes. In this regard let it be noted that copper will soften significantly at temperatures in the region of 150 °C. The softening temperature of a metal can be affected strongly by very small amounts of another metal, so firestop-pipe integral design is possible.

Castleford High School

An arson attack on this school in 2007 was the second such attack on the school. Three teenagers were arrested in connection with the fire.

Catalytic converters in cars, as a source of fire

A catalytic converter removes unburnt hydrocarbon from the exhaust gases, converting it to carbon dioxide and water. The catalysis enables such conversion to occur at lower temperatures and lower concentrations of hydrocarbon than would be possible in the absence of a catalyst. Nevertheless, if a car is parked on a grassy surface, the temperature of the casing of a catalytic converter at the stage where the engine is turned off can be high enough to ignite the grass.

Catalytic gas sensor

A.k.a. a pellistor (*pell*etised res*istor*), such a device uses catalytic combustion of a flammable gas as the principle of operation. It comprises a platinum heating coil inside a ceramic pellet, the surface of which is coated with a catalyst. Contact of a flammable gas with the heated catalyst surface causes heating of the assembly and the rise in electrical resistance is detected and forms the basis of the signal. In practice two resistors, identical except for the presence of the catalyst on one and its absence from the other, are incorporated into the sensor, the difference in resistance of the two is compared. This cancels out any effects other than those due to the flammable gas being detected. (*See also* **200N CiTipel®**.)

Caulk & Walk®

Acrylic-based sealant material displaying **intumescence**, suitable for use with **CVPC** sprinkler piping. It is manufactured by Tremco whose HQ are in Beachwood, OH.

Centralia, PA

The scene of a mine fire which has been burning since 1962. There are two theories as to how it started. One is that it began with what was meant to be controlled burning of an abandoned mine, extinguishment having been incomplete, enabling combustion to redevelop. The other is that hot ashes had been dumped at the site and that they had provided an ignition source. Over the next several years there were a number of unsuccessful attempts to extinguish the blaze. Matters became critical when it was discovered that gasoline at a local filling station had, during storage in an underground tank, risen in temperature to over 75°C after the fire had spread in its direction. In 1984 Congress voted money to help the residents of Centralia to relocate. A few residents have remained to the present time. The fire is expected to continue for another 200 years or more.

Chafing pad

Fire hose is susceptible to damage to its outer jacket when being dragged along the ground. Hose which has been taken out of service on account of its age and condition can be used to make a chafing pad, which is wrapped around a hose which *is* in service, to protect it from the effects of abrasion.

Changi International Airport, ARFF at

This airport in Singapore, which opened in the early 1980s replacing a previous "Singapore Airport", has two fire stations for the runways and a smaller one for the terminals. Being close to the sea, it also has facilities for sea rescue. The three fire stations between them can accommodate 23 fire appliances. There is a Casualty Collection Station (CCS) where *triage* – arranging casualties in order of urgency of treatment – can be carried out. There is helicopter access to the CCS.

Chartek 7

One of a family of intumescent fire retardants made from epoxy resin. A patented product, Chartek 7 is manufactured by Azko Nobel and is widely used at offshore oil and gas installations where to its role as a fire retardant it adds that of corrosion protector. Other products from the same manufacturer include Chartek 3, one application of which is the coating of LPG vessels to protect them in the event of a nearby fire and, perhaps the most interesting of all, Chartek 59 which was the first such product to be made from epoxy resin. It was developed from a similar substance used in the Apollo spacecraft, where it had a role in protecting a spacecraft from the conversion of kinetic energy to heat on re-entry to the atmosphere. It was introduced into the oil and gas industry in 1974.

Chicago Fire department, use of PPV by

Chicago became the first major US city to use **PPV** in fire fighting when its fire department took delivery of a Mobile Ventilation Unit (MVU) in April 2005. The Department uses it for fires in tunnels and in high-rise buildings. The MVU has a Ford chassis and cab with a hook lift. The demonstration and trial of this apparatus when delivered to Chicago was a seminal event in fire fighting. Fire services in New York and in Tokyo used it as a basis for evaluation and possible introduction of PPV.

Chicago Hilton

Rapid extinguishment of a fire at this hotel in 2006 has been attributed to the sprinkler system. The fire, which began in a linen container, had been put out by the sprinklers by the time the fire department arrived. Smoke damage was light.

Chimney nozzle

Term applied to a nozzle for insertion into an inaccessible part of a building where there is fire: apart from a chimney, a false roof is the most obvious example. To facilitate access, such a device will be long (up to about a metre) and thin (typically 2–3 cm in diameter). Typical performance is 10 gallons per minute. A nozzle for use in "awkward" places having a sharply pointed tip to aid penetration is called a *piercing nozzle*.

Chirk

This town near Wrexham, Wales was the scene in late 2007 of a fire at a Cadbury's Chocolate factory, believed to have begun in a silo of cocoa. Cocoa behaves similarly to sawdust in combustion, being capable of flaming or smouldering combustion according to conditions. In 1985 in the US there was a serious fire at a chocolate manufacturing and storage facility belonging to the manufacturer Van Duyn. A large quantity of chocolate products ready for distribution was destroyed. (*See also* **Aarhus**.)

Chlorez®

Trade name (in the US: in Europe it is marketed as Hordaresin®) of a solid material comprising chlorinated alkanes. It is suitable for blending with substances including adhesives, foams, plastics and waxes. Chlorez® can be supplied in powder or in flake form and, by reason of its high chlorine content (> 70%), imparts flame retardancy to the material into which it has been incorporated. Chlorez® has no metallic content, a feature it shares with **melamine**.

Chlorotrifluoroethylene (CTFE)

Base stock for a totally non-flammable hydraulic fluid for use in aircraft and military vehicles. Its suitability for hydraulic aircraft braking systems has been emphasised, as conventional hydraulic fluid can be ignited through brake overheating in an unscheduled or non-standard landing.

Chocolate lily

Botanical name *Fritillaria camschatcensis*. It occurs in northern North America as well as in other parts of the world including Japan and Siberia. It is used as a **firewise plant** in Alaska.

Chokecherry

Botanical name *Prunus virginiana*. North American tree widely occurring over many states, especially to the east. Growing to heights of up to 25 feet, it is a **firewise plant** by reason of high moisture and recommended for use as such in several states.

Cinema fire in India, 1997

This fire in Dehli killed 59 people. It started at an electrical transformer. The cinema was full at the time of the fire, and passages to exits were blocked. Most of the victims had been in the balcony and died from smoke inhalation. Twelve persons, including the two brothers who owned the cinema, received prison sentences.

Circleville, OH

There have been complaints by residents of Circleville that the siren at the Fire Department's base is unacceptably loud. There have been measurements of up to 130 bB. Exemption of fire departments from noise control has been invoked in defence. A report from a different US state (Oregon) states that 130 dB can be safely withstood for one minute.

Circuit-breaker installations, fire protection of

An effective way to fire protect these is for a layer of intumescent material to be placed between the panel supporting the circuit breakers and the wall behind it. Intumescent substances for this application have been developed which on heating will fill the entire enclosure ("fuse box" in older terminology) containing the circuit breakers and seal off the opening to it.

Cirrus Building

Apartment block in Helsinki, the highest apartment building in Finland. A recent fire there, originating in a sauna, was prevented from escalating by the opening of a **Hi-Fog®** nozzle in the part of the sauna where the fire had begun.

City of Montreal

Steam ship which, whilst carrying cotton from New York to Liverpool in 1887, caught fire off the Newfoundland coast. It also carried a significant number of passengers. The vessel was totally destroyed. Occupants dispersed on to ship's boats and most were rescued, initially by a German barque carrying turpentine. They were then transferred to the steamer *York City*, which was on its way to London. Records from the period reveal that the incident was the seventy-third cotton carrying vessel to have caught fire within five months.

Claridge's Hotel, fire in 2004

This most prestigious of hotels in London's West End was the scene in December 2004 of a fire which necessitated evacuation of the 1300 persons inside. There were, as would be expected, high-profile persons staying there at the time and press conferences with some of them scheduled for later in the day had to be cancelled. The fire began in the kitchen. There were no injuries.

Class F Fire

Fire involving cooking oils or fats. Sometimes fires with these materials have been treated as Class B, but extinguishing agents for such are inadequate when applied to a fire having originated with a cooking oil or fat. **"Wet chemical" extinguishing agents** have been developed for these. Some trade literature still describes fires involving fat or cooking oil as Class B, which is not incorrect if the view (which the author has encountered) that Class F is a sub-class of Class B is taken.

Class K fire

Same as Class F. Class K is the North American term for a fire involving fat or cooking oil, termed elsewhere a **Class F fire**.

Clean-up gloves

Gloves for fire fighting, such as the Blazemaster®, are expensive and, though of robust construction, do wear out sooner or later. Once a fire is extinguished there is still work for the fire fighter to do. In order to extend the life of the fire fighting gloves, as well as for his own comfort, he might at this stage change into clean-up gloves, a.k.a. utility gloves. The function of these is to afford protection from sharp objects and abrasion. Cow hide is a possible material for their fabrication and this in suitable thickness does provide moderate thermal protection in the event that a fire fighter unexpectedly picks up something hot. The synthetic material Clarino is also used for clean-up gloves.

Clear Creek County, CO

The scene in October 2007 of a fire during maintenance at a hydroelectric plant in which five workers died. They were in an underground tunnel spraying epoxy sealant on to its walls when the fire broke out. The workers were wearing protective apparel made of Tyvek®. This material, developed by Dupont, is very effective in protection against chemicals. It is not however fire resistant: on the contrary, the Materials Safety Data Sheet on Tyvek® strongly deprecates its use in fireproof clothing.

Clearwater County, MN

Scene of a recent pipeline fire in which two persons, both young maintenance engineers, died. Crude oil from the pipeline exited under pressure and ignited, resulting in an explosion with overpressure. The pipeline was of 34 inch diameter and had received a temporary repair for a leak two or three weeks earlier. The pipeline is part of a network which conveys crude oil from Canada for refining.

Cleveland School, Kershaw County, SC, fire at in 1923

The death toll was 77, some bodies never having been identified, when the Cleveland School burnt down on 17 May 1923. An illuminating lamp (probably using kerosene, which was widely available for such use in the US long before 1923) fell from its installation on a ceiling on to the platform below, where straw and other ignitable materials were present as part of the décor for a school play. Even graver was the school fire at **Collinwood, OH** 15 years earlier. (See also **Our Lady of Angels School**.)

Coastal redwood

Botanical name *Sequoia sempervirens*, an example of a "**highly flammable tree**". It occurs along the Pacific coast of North America.

Coatbridge, Lanarkshire

This town in south west Scotland was the scene in January 2008 of a fire involving 400 tonnes of tyres. Overhead power lines for the railways were damaged and disruption to services resulted. Tyres tend to burn very smokily as was observed in the fire under discussion. (See also **Rochford, UK**; **Watertown, WI**.)

Cocoanut Grove[9]

Name of a nightclub in Boston, MA where, in November 1942, there was a fire which caused 492 deaths. The cause of the fire was ignition of thin paper adornments within the club building. Amongst the difficulties with

---

9 An article to commemorate the 65th anniversary of the Cocoanut Grove fire appeared in the November/December 2007 issue of the *NFPA Journal*.

escape was the fact that the exit doors swung open in a direction opposite to that of movement of persons.

Colac

Rural town in Victoria, Australia, the scene in January 2008 of a fire at a timber mill which resulted in damage worth A$3 million. The fire, which was attended by over 40 fire fighters, is thought to have begun during the process of drying timber in a kiln.

Colet Jaguar

Range of **ARFF** vehicles produced by Colet SDV in Newark, CA. Across the range, structural parts of the vehicle are made from stainless steel and panelling from aluminium. The smallest in the range, coded the K/15, is a two-axle vehicle with a 600 hp diesel engine capable of carrying 1500 to 2000 gallons of water. The next up, the K/30, has three axles and can carry up to 3000 gallons of water. The top of the range is the K/R40 with five axles and a 1600 hp engine. This is one of the most advanced ARFF vehicles available. It can accelerate from 0 to 50 mph in 17 seconds, can release water at 1500 gpm via a **turret** and has a foam proportioner. In the early days of its manufacture, the Jaguar proved its worth by a remarkable performance during a fire at **Hartsfield International Airport**. The **Snozzle™** has sometimes been retrofitted to Colet ARFF vehicles.

Collinwood, OH

The scene in March 1908 of a school[10] fire which claimed 174 lives: 171 pupils, two teachers and one rescuer. A steam pipe caused a wooden support structure to heat and ignite. Propagation was accelerated by flammable floor coatings. There was a political spin-off from the fire: Collinwood surrendered its autonomy when it declared itself unable adequately to provide for fire safety and became part of Cleveland, OH.

Colorado, Department of Public Safety

The term "Fire Marshal" means different things in different states of the US according to local statutes. However, all but two of the lower 48 states have a Fire Marshal: the exceptions are New York (as discussed in another entry) and Colorado. The Colorado Department of Public Safety has a Fire Safety Division, the head of which reports to the departmental head. The divisional head represents the State at **NASFM**.

10　The Lake View School. Its replacement was called the Collinwood Memorial Elementary School.

Combination stream

Water release from a nozzle such that there is straight stream flow in the centre surrounded by fog at a wide **fog angle**. Nozzles are available which will provide such a pattern and they include one in the ProJet range. One nozzle type capable of producing a combination stream is the **Saberjet™**.

Compressed Air Foams (CAF)

These comprise a foam solution, such as might be used conventionally in fire fighting, through which air under pressure has been passed with a resulting expansion of the foam. Though such systems were first tried out in the 1930s, it is only since the 1990s that there has been serious R&D into them and use on a trial basis by some fire departments. Advantages of CAF over plain water or foam without the treatment with air include the following. A foam so produced is significantly less dense than water, making a hose bearing it lighter than one bearing an equivalent volumetric flow of water. Its lower density also increases the distance travelled on ejection from a nozzle, which means that fire fighters can, other things being equal, stand a distance equivalent to that increase further away from the fire. The surface tension of such a foam causes it to remain in contact with a vertical wall or a ceiling to which it has been applied, avoiding the wastage of heat uptake potential when water simply runs off such surfaces. In enclosure fires this could quite conceivably prevent flashover.

The R&D referred to has been partly into "knockdown times", that is the time taken in a simulated enclosure fire for the temperature to drop from 1000°F (538°C) to 212°F (100°C). Quite dramatic results in such tests have been reported, for example temperature drop rates averaged over that interval of 3.5°F s^{-1} with water alone, 7.6°F s^{-1} with water containing a foam and 20.5°F s^{-1} with CAF. There is much interest in changing from water to CAF in routine fire fighting and at the present time fire departments in countries including the US, the UK and Germany are involved in field trials. In places including Tyne and Wear, UK the community has been consulted about proposals whereby fire brigades will replace simple water extinguishment with CAF systems. Informed opinion was strongly in favour of such replacement. The **aspirated foam nozzle** works along similar principles to CAF.

Constellation aircraft, crash involving in 1975

By 1975 the Lockheed Constellation had of course long been taken out of passenger service but some remained in use in **air tanker** service, as with for example the **DC-4**. A "Connie" crashed in Arizona in May 1975 with loss of several lives. It had caught fire shortly after take-off, when black smoke was seen coming from it. Piston engine aircraft require anti-detonation injection fluid (ADI), containing methanol, on take-off and a difficulty with this operation is believed to have been the cause of the fire.

Contaminated fire water, environmental hazards of

Water from fire extinguishment is likely to contain residues of halogenated foaming agents and of course contaminants including suspended particles from the fire itself. At a large industrial site where, in the event of a major fire, amounts of contaminated fire water would be very large, certain measures are possible. Sometimes these are part of the system in place at the site to contain a liquid spill from a storage tank: a simple bund put in position for this purpose can double up as a very effective restraint on contaminated water spread. Otherwise a lagoon can serve this purpose. This might be a "sacrificial area" at the site but will not necessarily be: sometimes a part of the site, for example a car park, might be deliberately isolated from the water drainage system so that it can be used as a "lagoon" if necessary. There are also tanks, installed or portable, for containment of contaminated fire water. Measures such as lagoons and bunds will not of course totally restrain fire water, and after a major fire there is a need to monitor watercourses into which the fire water might have found its way, as in now taking place in Hertfordshire, England after the 2005 Buncefield accident. Most of the extinguishing water at Buncefield was, however, retained and has been put into storage containers in the custody of organisations including Thames Water. In smaller scale fires it is possible to prevent influx of fire water into drains by use of clay mats or a putty-like material. The London Fire Brigade practise this. (*See also* **perfFECT™.**)

Contender

Series of appliances for water pumping ("pumper", or "pumper-tanker") from Pierce Manufacturing. It is, as such devices go, fairly small, having a maximum water delivery rate of 1250 gpm. The entire construction is on quite a small scale, making for a short wheelbase, resulting in good manoeuvreability. One of the most recent deliveries of a Contender was to Blair County, PA. This example has a pump made by Waterous. About a year earlier three new Contenders were delivered to Sumter County, FL each of which has a Hale pump. Delivery of water with either pump is up to 1250 gpm. Pierce Manufacturing also build larger pumpers than the Contender, for example a Pierce apparatus with 2000 gpm pumping rate and a tank for 1800 gallons of water was recently delivered to the Fire Department at New Wilmington, PA. On the other hand Pierce produce a smaller pumper than the Contender, the Pierce Mini-Pumper, which, in successively improved versions, has been obtainable for over 30 years. A new Mini-Pumper was acquired recently by the Fire Department in Lyons, Wayne County, NY. It can deliver up to 500 gpm of water and can store up to 300 gallons. It is built on a Ford 4WD cab/chassis unit shared with certain models available generally to commercial or private owners. A Contender to which no pump has been fitted can, according to customer's specifications, be set up as a **rescue vehicle**.

Contra Costa County

The Fire Protection District bearing the name of this Californian location recently took delivery of a new **quint** built by American LaFrance. The reach of its ladders is 100 feet and its pump, manufactured by Hale, can deliver 2000 gpm of water. **American LaFrance**, with its HQ in South Carolina, has for a long time been a notable producer of quints amongst other sorts of fire fighting appliance. In the early post-war years an American LaFrance quint was available which had a ladder reach of 75 feet and a pumping capacity of 750 gpm of water.

Controlled burning

A.k.a. back burning, the intentional destruction by fire of vegetation to reduce the amount of fuel present in the event of accidental fire. It is usually from the ground: if it is from the air it will involve the **helitorch** or the "**ping pong ball**".

Convair 580

A passenger aircraft having two turboprops manufactured several decades ago, many of these were converted into air tankers and about ten are still in such use in Canada. These have a water holding capacity of about 2000 gallons and foam proportioning facilities.

Convergency

In fire fighting the term means random assembly of officials, voluntary helpers and onlookers at the scene of a fire, which can inhibit the work of professional rescuers. The word was used in a follow-up to the fire involving a Boeing 737 aircraft at Manchester Airport in 1985 in which there were 55 deaths. Of the categories of persons so converging, press representatives were selected for "dishonourable mention", not only at the scene of the fire but at the hospitals.

Coolidge dormitory

This dormitory at the University of Massachusetts in Amherst was in February 2006 the scene of a fire which began with an unattended candle. The dorm was fitted with a sprinkler system and this mitigated the consequences. Several student occupants did not evacuate the building when the alarm sounded, and this is in violation of regulations. The matter was referred to the Dean of Students for possible disciplinary follow-up, as was the lighting of the candle which caused the fire.

Co-op supermarkets

The Co-op Group was fined £250 000 after an inspection in 2006 of its supermarkets in the East Sussex area revealed fire protection difficulties in some of them. Such difficulties included locked fire exits, fire doors wedged open and too few alarms.

Copleston High School

This school in Ipswich, England was damaged by an intentionally lit fire in August 2006. An 18-year old male was arrested in connection with the fire.

Copper compounds, use of as flame retardants and smoke suppressants

There are many proprietary flame retardants and smoke suppressants for use with polymer materials, some of which have been the subject of patents. Compounds of copper, including copper(II) bromide and copper(II) oxide, often occur in such retardants and suppressants. Notable amongst copper compounds so employed is copper oxalate, CuC_2O_4, which is widely used in the fire protection of PVC. Copper stearate and copper oleate, huge quantities of each of which are produced for paint and varnish manufacture, have also featured in patents for flame retardants and fire suppressants.

Corby, Northamptonshire

The scene in May 2008 of a fire at an adhesives factory believed to have begun with the ignition of a vapour by an electrostatically generated spark. It was attended by 60 fire fighters and two persons were hospitalised for smoke inhalation. It was feared that phosgene might have been amongst the combustion products. (*See also* **Podolsk**.)

Coronado National Forest

This forest, which is partly in Arizona and partly in New Mexico, was the scene in July 2007 of a fire which was initiated by lightning. An area in excess of 23000 acres was affected and there were 13 fire fighter injuries. (*See also* **USA, lightning fires in**.)

Coryton refinery

The Coryton refinery, east of London, has been in operation since 1953. It was originally a BP facility and is now one of six refineries in Europe operated by the Swiss concern Petroplus AG. The present refining capacity is about 65 million barrels per year and the refinery has the space to store 4 million barrels of crude. The Thames estuary location provides for supertanker access. On 31 October 2007 there was a fire at the refinery. Early reports stated that the fire started in a naphtha column. Naphtha is the second "lightest" distillate after gasoline.

Cosmo Oil Company

A desulphurisation plant near Tokyo belonging to this company was recently closed down after an explosion. Refining facilities at the same site, of capacity 240000 barrels per day, were not affected.

Cospatrick

Sailing ship which, whilst taking migrants from England to New Zealand in 1875, caught fire. The death toll was in excess of 450. She was constructed of teak and had previously been used as a troop carrier between England and India. A sailing ship required time of the order of months to make the journey from England to the Antipodes, and the *Cospatrick* was over two months into the journey, fairly close to the Cape of Good Hope, when fire broke out. The vulnerability of the vessel in a fire due to its being made of wood was exacerbated by the fact that coal tar – an effective fire accelerant – was also amongst the construction materials. The only survivors were those who, having departed in the ship's boats, were picked up by the vessel *British Sceptre* en route to St Helena.

Cotton fabrics, fire retardants for

Cotton consists of cellulose, which readily burns with a heat of combustion of about 16 MJ kg^{-1}. The organic chemist W.H. Perkin (1860–1929), who had extensive involvement with the development of synthetic dyes to replace natural ones, was the developer of the first commercial fire retardant for cotton. This was simply stannic chloride (tin(IV) chloride in modern terminology). Perkin developed what he called the "Non-flam", process whereby this substance could be incorporated into a cotton garment sufficiently tenaciously for it to withstand two years of weekly washing without loss of the retarding effect. Tin(IV) chloride owes its effectiveness as a fire retardant to its ability to release chorine atoms, which can scavenge reactive intermediates in incipient combustion, especially hydrogen atoms.

Whereas Perkin's approach was incorporation of an intrusive chemical substance into the fabric, retardants were later developed which reacted chemically with the cellulose at the small depths into the cotton fibres at which oxygen uptake was sufficient for significant reaction. A family of substances called tetrakis(hydroxymethyl)phosphonium salts has been developed as such retardants for cotton and other fabric materials including polyesters. An example is tetrakis(hydroxymethyl)phosphonium chloride (THPC):

$$[P(C_2H_4)_4]^+ Cl^-$$

molecular weight 190.5. Other tetrakis(hydroxymethyl)phosphonium salts used in fire retardation include the sulphate and the acetate. In particular, prior treatment of THPC with urea under conditions of carefully controlled pH produces "THPC-urea". First reported in 1974, THPC-urea is now the most widely used of the retardants based on tetrakis(hydroxymethyl) phosphonium salts, accounting for about half the total world production at the present time.

The most recent R&D into fire retardation of cotton has been into the use of nanoparticles, that is particles with diameters 100 nm or less. Such particles composed of clay or of zinc oxide have been investigated as fire retardants. Application of these to a fabric affects characteristics other than fire susceptibility, for example how water repellent it is. **Borax** is also applied to cotton, especially in mattresses.

Cotton fires, examples of

Cotton is composed of cellulose and burns like paper. There have been many fires over the years in the cotton industry, including one in Burnley, England in January 2008, which is believed to have originated with over-heating of a piece of machinery. In Gippsland, Australia just over two years earlier, fire at a cotton spinning mill caused damage worth A$0.5 million. Moving further back in time, large amounts of cotton were present at the **Triangle shirtwaist factory fire**. Moving further back still, in late 19th century America fires involving cotton on ships, or in sheds at harbour cities awaiting loading on to a ship, were legion. The **City of Montreal** is a case in point.

"Courage" Glove

Glove for fire fighter use having, in common with many others including the **Blazemaster®**, a Nomex®/Kevlar lining. It is manufactured by Fire-Dex in Medina, OH. Its rare feature is goat hide as the material for the palm of the glove.

CP-680

Range of **cast-in-place firestops** manufactured by Hilti in Tulsa, OK. The CP-680-P range is for both polymeric and metallic pipes and the CP-680-P-M series for metallic pipes only. Each comes in a range of diameters, and accessories include extensions for applications where the concrete is deeper than usual.

Crested wheatgrass

Botanical name *Agropyron cristatum,* a grass which tends to resist fire spread because of its growth structure, which provides for a high thermal inertia. It is used as a **firewise plant** in the US, notably in Utah.

Crimson

US manufacturer of fire appliances based in South Dakota offering a range of aerials, pumpers and tankers. The parent company is Spartan Motors and the Spartan name is sometimes used, for example the Spartan Gladiator, which amongst other uses can form the cab of a **tractor-drawn aerial**.

Crosstech®

Material suitable for use as a **moisture barrier** in fire fighting apparel, developed and manufactured by W.L. Gore and Associates whose HQ is in Delaware, USA. It has the intrinsic permeability to vapour and resistance to liquid water ingress required of a moisture barrier (which in that sense acts as a membrane). Crosstech® has satisfactory **THL** and thermal protection characteristics. As well as being incorporated into multi-layer fire fighter protective apparel Crosstech® has been used to line fire fighting boots and has been used in hand protection, for example in **Eska fire fighting gloves**. A similar product to Crosstech® at a budget price is **RT700A®**.

Crowd suffocation

A.k.a. compressive asphyxia, a fatal factor in crowd disasters, including those arising from the need to escape from a fire. Probably many deaths attributed to physical "crushing" in such incidents have in fact been due to this phenomenon. (*See also* **Paisley, Scotland**.)

Crown of the head, role of in fire fighter survival

A fire fighter needs to be fully protected by the most effective apparel available and will be heavily clad. However the body also needs to be able to transfer heat and such transfer is obviously restricted by heavy protective apparel. Sometimes in the fabrication of a **fire fighter hood** provision is made for this by incorporation of a ventilating panel to fit over the crown of the head. Heat can then be transferred from this region of the body. An example is the Morning Pride fire hood. The Morning Pride fire hood is made from **Nomex®**.

Crown Victoria Police Cruiser (CVPC)

Vehicle for police use in the US, manufactured by Ford. There having been a number of fires involving such vehicles through fuel loss on impact, demands have been made on Ford to recall and strengthen them. This will involve a "bladder" lining the tank, which will prevent spillage if the fuel tank is broken open. A further measure is installation of a plastic receptacle containing fire-retardant powder, which will be released if the tank contents escape. According to tests conducted in Utah a vehicle having a tank so protected can withstand a collision at 81.9 mph without a resulting fire.

CVPC

Chlorinated polyvinyl chloride. The monomer of PVC, formula $CH_2=CHCl$, is of course far from being fully substituted and it is possible to modify the properties of the polymer as required by further chlorination. Amongst the uses of CPVC in fire protection is piping for sprinkler systems. (See also **Sprinkler systems, antifreeze for**.)

D

Dagger

From Dennis, the manufacturer of the **Sabre**, fire truck cab/chassis combination subsequently suitable for building a **water tender** or a pumper. Like the Sabre it uses a Cummins diesel engine.

Dallas, TX, apartment fire 2007

Dallas saw its worst ever apartment fire in March 2007. Classified as 7 on the **alarm number** scale, the fire caused one death and 17 non-fatal injuries. Many apartments in the block were totally destroyed.

Danvers, MA

Seventeen miles from Boston, the scene of a recent explosion at a paint and ink factory. Although there were no deaths or serious injuries, many buildings were badly damaged and their occupants rendered homeless. It has been suggested that leaked natural gas might have had a role in the accident.

Darby Township, PA

Good Will Fire Company # 1 at Darby numbers amongst its appliances a **water ladder** manufactured by **Crimson** and acquired new by the Company. Its pump, capable of delivering 1250 gpm, is a Hale product and the water tank will hold 650 gallons. It also carries **PPV** facilities.

DC-7

This was the last piston engine aircraft from the Douglas Corporation, the DC-8 and later being jets. Many DC-6 and DC-7 planes were converted to **air tankers** and at the time of writing four DC-7s are still in such use. Each has a capacity of about 3000 gallons of water.

Darley 800

Double-jacket fire hose. The jacket weaving is so fine that water penetration is precluded and no drying is required after use. The inner lining is synthetic rubber. It is intended for use with **attack apparatus** and comes with a ten year warranty. The "800" in the trade name denotes pressure testing to 800 psi. The same convention applies to **Tidalwave 600™** hose.

DarQuest Lightweight

Fire hose having an outer cover containing **Hypalon®**. When Hypalon® was developed by DuPont, it was enthusiastically received by hose manufacturers, although its prohibitive price meant that grades containing significant amounts of fillers and extenders were in fact present in the product as supplied for hose manufacture. DarQuest has a woven jacket and a thermoplastic liner and is suitable for use in attack.

Daru

Port in Papua New Guinea where in January 2007 there was a fire involving a vessel with a payload of wood products, including veneer and plywood. The crew abandoned the vessel and a tug having **FiFi I** classification was sent for and arrived about 24 hours later. Three days on the vessel were needed to fight the fire but it had to be released for other previously made commitments. It was replaced by the *Pacific Responder*, an emergency towing vessel registered to Australia which also had FiFi I classification. This was the first response of the *Pacific Responder* to a genuine emergency.

Dayton, OH

The scene in 1987 of a fire at an automotive paints warehouse. Flammable liquids in a total quantity of 1.5 million gallons were stored in containers of size up to 5 gallons. Smaller containers, some only one pint in volume, were held in cardboard cartons, ten of which fell on to the floor during a forklift truck operation. There was container breakage, leakage, ignition and rapid involvement of other containers, which on heating blew off their lids and added fuel to the fire.

Decibel (dB)

Logarithmic scale for intensity of sound. In assessing the magnitude of sound on this basis the lowest level of sound audible to someone with healthy hearing is assigned value p_0 and the actual sound of interest p. The sound in dB is then:

$$dB = 20 \log (p/p_0)$$

so p_0 corresponds to zero on the decibel scale but that does NOT signify silence. If the level is doubled, the value in decibels goes to:

$$20 \log 2 = 6$$

or if it goes up by 100:

$$20 \log 100 = 40$$

Decibel values higher than about 80, if experienced for long periods, can cause damage to the hearing. The decibel scale is important in fire engineering in that it is the measure upon which alarms and sirens are assessed. (See also **Banshee™; Circleville, OH; Kidde 0915Uk; Sounders, for fire alarms.**)

Deep snow fire hydrant

Often in areas where snow drifts occur fire fighters rely on prior knowledge of the location of a hydrant and its vivid colour in order to locate it. There are, however, fire hydrants which have a height of 6 feet or more above the ground (meaning that the lengths above and below ground will be comparable). Such are in use in Niigata, Japan.

Delhi India, fatal fire in January 2008

Standards of fire protection and of fire fighting are lamentable in India and fatal fires are numerous. At the time that this book was being prepared there was a fire in the Delhi slums which claimed at least six lives and caused critical injuries to ten others. Three hundred homes (such as they are) were destroyed in the fire to which 30 fire appliances from around Delhi had been sent .

Delta™

Material for fire fighting apparel, made by Dupont. It forms an outer layer, on top of any moisture barrier. Its composition is Nomex® 75%, Kevlar® 23% and **P140** 2%·

Delta Hotel

This building in San Francisco was a "hotel" for legal purposes including fire protection, but had more the nature of a hostel for persons on low incomes. Amenities were basic and did not extend to sprinklers. It burnt down in August 1997 with the loss of six lives. This tragedy was one factor in the campaign which led to the **Residential Hotel Sprinkler Ordinance, San Francisco**.

Denver, CO, death of a fire fighter May 2006

A 61-year old fire fighter "15 shifts shy of retirement"[11] experienced **cardiac arrest** whilst attending a building fire in the Denver area and became trapped within it. He had entered a bedroom and was found unconscious there by a fellow fire fighter who took him outside. He was not breathing at that stage. He had no burn injuries. His death a week later followed his removal from a life support system with the family's consent.

Detention fire equipment

Facilities for installation of fire fighting equipment for immediate use when required. It includes hose reel cabinets and fire extinguisher holders. A holder for a fire extinguisher might be simply a metal frame, or it might be a moulded plastic structure: with the latter, holders for pairs of extinguishers are common. There are also cabinets for fire extinguishers, necessary for outdoor conditions or where an extinguisher is carried by a vehicle. These are often made of plastic. Such a cabinet can be front loading or top loading. An extinguisher from one manufacturer and detention equipment from another can be "mixed" provided that size and weight limitations of the detention equipment are not exceeded. We should note that it is a British Standards requirement that a fire extinguisher be either contained or supported and not be simply standing on a floor. If the support has wheels and handles, the extinguisher and support comprise a **trolley unit**.

Deutsche Bank Tower, New York

Two fire fighters lost their lives during a fire at this already partially demolished building in August 2007. It has been suggested that a **standpipe** failed and in the follow-up the matter of responsibility for periodic checks of these was raised. The FDNY have stated that it is required to inspect them only every five years and that the building owner has responsibility for checks and maintenance between visits by the Fire Department.

Dhyaneshwar Nagar

Slum area of Mumbai, India where seven people including a number of children were killed in a fire in November 2001. Several other people were injured. The vulnerability of bad housing to fire was exacerbated by fire fighting standards below those which would prevail in a more advanced country. There was also evidence of this at **Kidderpore**.

---

11 *Rocky Mountain News.*

Diamond Princess

This luxury vessel was the scene of a major fire in 2002 whilst she was still being built at the Mitsubishi shipyard in Nagasaki. At the same time a sister vessel the *Sapphire Princess* was being built at the same shipyard. This took the identity of the *Diamond Princess* and the hull of the original *Diamond Princess* became that of the *Sapphire Princess*. Both vessels duly entered service. (*See also* **Genesis**.)

Diffuser, use of in fire extinguishment

A possible alternative to a conventional **sprinkler head** is a diffuser, which works as a nozzle in reverse. In such an application a diffuser would receive water at high velocity and decelerate it for focused application. The calculation in the boxed area below explains the principles.

Imagine that water exiting a supply at 1 m s^{-1} has to be decelerated to 2.4 cm s^{-1} for fire protection purposes. If it is received into a horizontal diffuser of 5 cm diameter, calculate the diameter to which the diffuser must extend in order to achieve the required delivery speed.

The statement that the diffuser is horizontal is a cue that potential energy effects are nil. In orientations other than horizontal, this is not necessarily true for liquid flow.

Letting subscript 1 denote entry and subscript 2 denote exit, continuity gives:

$$c_1 A_1 = c_2 A_2$$

since the flow is incompressible.

$$A_2 = A_1 (c_1/c_2)$$

$$d_2 = d_1 \sqrt{(c_1/c_2)}$$

$$\Downarrow$$

$$d_2 = \underline{33 \text{ cm}}$$

Diktron DSX

Automatic distress signal unit (ADSU) of UK manufacture and used by some UK fire services. It is designed to be incorporated with a **SCBA**. It releases sound in the frequency range 2.7 to 3 kHz and also has a light which flashes in a "man down" situation. It is powered by a 9 V battery.

Disflamoll®

Family of phosphorus-containing flame retardants produced by Lanxness (formerly part of Bayer), which has its HQ in Pittsburgh. Disflamoll® TKP and Disflamoll® TP are respectively **tricresyl phosphate** and **triphenyl phosphate**. Disflamoll® DPK is diphenyl cresyl phosphate, a dual plasticiser and flame retardant for such materials as nitryl rubber and PVC. Disflamoll® DPO is diphenyl-2-ethylhexyl phosphate, which is also produced in large quantities in China. Two other groups of flame retardants come from Lanxness: Levagard®, developed for polyurethane foams, and Bayfomox®, which is for use in buildings and displays **intumescence**. Levagard® products contain phosphorus.

District of Columbia, anomaly in the fire death rate of

DC has a fire death rate (2004 reckoning) of 36.1 deaths per million population, higher than that of any of the states and almost three times the national figure of 12.4 deaths per million population. DC has boundaries with Maryland and with Virginia, the fire death rates of which for the same year are respectively 15.8 and 13.1 deaths per million population. Like New York City, DC is not rated under the **Public Protection Classification (PPC™)**. The value for New York is fairly close to the national figure, making that for DC all the more surprising.

Dixmude, Belgium

The scene in 1933 of an air crash originating with an on-board fire. The aircraft was an Argosy operated by Imperial Airways and it had taken off from Brussels for London. All on board were killed. It was believed, though never proven, that the fire was lit intentionally by a passenger, in which case the incident was probably the first example of sabotage of a passenger aircraft in flight.

Dogs, use of by fire services

In the days when fire apparatus was horse drawn, such apparatus would often be led to its destination by a Dalmatian dog. Once at the fire they would guard the horses. One reason for the choice of the Dalmatian was that its appearance was, to a horse, distinguishable from that of another horse. In today's world dogs are used by some fire services to find trapped victims in a fire and to locate dead bodies. In the UK there are 13 dog rescue teams, each managed by fire fighting personnel from a particular brigade and operating on a roster basis. Such a team was sent to Glasgow in 2004 when there was a severe fire at a plastics factory which claimed nine lives. No particular breed of dog has been identified as being the most suitable for training in rescue and there were more than 15 breeds amongst those deployed at the WTC after 9/11, as well as some of no pedigree at all.

Dokkhamtai

Province of Thailand, the recipient of a second-hand fire truck with a pump from Japan under its Grant Assistance for Grassroots Human Security Projects Scheme. The funds allocated for its transportation from Japan were US$14 555. The appliance was badly needed. Significant infrastructure by way of hydrants and the like will be needed if the appliance is to function to full effect.

Domestic chimneys, extinction of fire in

In such a fire, application of water directly from a hose–nozzle system using the sort of hardware examples which have been discussed in this volume might be overkill and lead to excessive water damage. In such cases regular garden hose can be used, having been connected to the stronger hose by a suitable adapter. It will have a disperser at the end which will lead to a spray hence to maximum return in terms of extinguishment on the water applied. Some fire services refer to such a device as a "master fire set".

Don Inda

Coastal protection vessel for Spanish waters, entering service in late 2006, equipped for fire fighting at the quite advanced level of **FiFi II**. Additionally, like many other such vessels, including the **Anglegarth**, *Don Inda* has a supply of water to protect the vessel itself when on fire fighting duty. The sister ship the *Clara Campoamor* was delivered in March 2007. Like the **UT 512-OPV** which serves in Norwegian waters and is rated FiFi I, *Don Inda* and *Clara Camoamor* were designed by Rolls Royce Marine as was the **Abeille Bourbon**, which operates off France.

Dormitory safety, enhancement of by sprinklers

This has been the subject of debate and discussion very widely, and also of legislation. Convincing evidence of the efficacy of sprinklers in "dorms" is provided by a 2005 study conducted jointly by the US Fire Administration

and NIST in Washington. Two simulated fires, identical in their initial conditions except for the presence of sprinklers in one and their absence from
the other, were examined. Without the sprinklers, temperatures of 120°C
were measured 23 m from the origin of the fire within three minutes. With
the sprinklers, a temperature of this value was not reached even much
closer to where the fire had begun. There have been very many tests of
this sort relating not only to student residences but also to buildings such
as barracks and department stores. The value of suitable sprinklers is
invariably clear in such investigations. The matter of sprinklers for buildings is not of course apolitical. The hoteliers of **Virginia Beach** on being
told that they had to install sprinklers attempted to make a case that
the same requirements were not being imposed on student dormitories
where at least in some cases the money would have had to be found from
the public purse. (See also **Coolidge dormitory**; **Elmhurst College**; **Fire
Sprinkler Tax Incentive Act**; **Hester Hall**.)

Douglas C-54G-DC

Military variant on the DC-4 passenger aircraft, having featured in the air
lift to Berlin. One or two survived into the 21st century as freighters for
carrying fuel to remote localities. In January 2007 such an aircraft, operated by Brooks Fuel in Fairbanks, AK, was taking fuel oil to Nenana, AK, an
isolated settlement with a population of a few hundred, when it crashed
because of an engine fire. The crew escaped injury and the payload of 300
gallons of fuel oil, held in "bladders" was salvaged. The aircraft, 62-years
old and in a sense irreplaceable, was lost.

Dowtherm A

Heat transfer fluid containing biphenyl and diphenyl oxide. Its flash point
is 113°C. This is a low flash point having regard to temperatures which a
heat transfer fluid will experience, so avoidance of vapour build-up during
use is necessary.

Dromadear

Small **air tanker**, capable of holding up to 400 gallons of water. That is a
sixtieth of the amount which the **Boeing 747 Evergreen Supertanker** can
carry and many helicopters for fire fighting use can carry much more than
the Dromadear can. Similar in water holding capacity to the Dromadear
is the **Turbine Thrush**.

Drum safety cabinet

Fireproof cabinet built to hold a drum of flammable liquid. Orientation
may be vertical or horizontal. In the horizontal case there will be a support

("cradle") for the drum. Typically, such a cabinet will hold a drum in the size range 30 to 55 gallons. In vertical orientation the cylinder can have a funnel at the top, and waste flammable solvent can be collected in this way. "Hybrid" cabinets, known as double-duty cabinets, are also available. These have a vertical drum in the 30 to 55 gallon range separated by a partition from a conventional cabinet holding small amounts of liquids in multiple containers.

Dry fire hydrant

For use in areas where water supply from pipes is lacking, a dry fire hydrant draws water for fire fighting from lakes and ponds. Unlike water in a city water supply, such water is not pressurised therefore a pump is needed to raise the water, which is also strained to remove suspended particles. In this way a fire fighting crew can replenish its water supply. Unlike a conventional fire hydrant, which is simply purchased and installed, compliance with relevant standards having been confirmed by the manufacturer, a particular dry fire hydrant has to be designed to fit in with the terrain close to the water source. **uPVC** is a common choice of material for pipe work in a dry fire hydrant.

Dual flow nozzle

Adjustable gallonage nozzle in which only two gpm ratings are possible. An example is the Gladiator® Dual-Flow Nozzle, which at 100 psi pressure releases at 1500 or 3000 gpm according to selection. It can be used with **AFFF** and also with AR-AFFF.

Dubai Tower

Once it is complete this will be the highest "free standing structure" in the world. In 2007 at least two construction workers at the Tower lost their lives in a fire. Many others had serious injuries. One of the deaths was due to a fall during an attempt to escape.

Duchess Anna Amalia Library

This valuable collection in Weimar, Germany[12] was partially lost as a result of a fire of electrical origin in 2004. In fact, 50 000 books were destroyed out of a total of 112 000 in the building and some of those not totally destroyed were nevertheless damaged. After the fire, volunteers moved the surviving books to safety. Many German business firms contributed to the expense of restoring the building and of reconditioning those books

---

12   Until the reunification of Germany in 1989 Weimar was in East Germany, a.k.a. the DDR.

which had been damaged. The building re-opened in 2007. Those books unaffected by the fire were put back into the building, whilst a program to restore the salvageable ones continues and will probably take until 2015. The re-opened building is protected by a **Hi-Fog**® system. One reason for the choice was that fine droplets make for a lower degree of non-thermal damage due to water. Happily, many of the books lost were not irreplaceable: there are other equally authentic copies in existence, which the library can acquire as funds become available.

Dulles Airport

A passenger area covered by a glass dome in the Airport Train System at Dulles Airport in Virginia is protected by a **Hi-Fog**® installation. A deciding factor in the choice was water availability, and the wide-diameter piping necessitated by high water supply rates. It was shown that whereas a particular conventional sprinkler system which was being assessed would have required 11 355 litres per minutes to protect the domed area, the **Hi-Fog**® device could protect it as effectively with 1363 litres per minute.

"Dump-and-run"

Term applied in rural fire fighting in particular to a procedure whereby a **water tender** vehicle offloads water into a **dump tank** for transfer to the **attack vehicle** and then departs to take on more water for subsequent transfer to the dump tank. Other such jargon includes "**Fire-Attack mode**", "**Gravity-Nurse**" and "**Pumping-Nurse**".

Dump tank

A.k.a. a relay tank, used to hold for short periods water brought by a **water tender** to await application to a fire, possibly after a foam concentrate has been added. Such a tank is made of flexible material with a metal folding frame and needs to be convertible from collapsed form to functional in a matter of minutes. Common liner materials are **Hypalon**® (chlorosulphonated polyethylene) and canvas. One commercial design of dump tank amongst very many is the Fol-Da-Tank™, some 40-year old examples of which are still in service! Exit of water from the **water tender** is via a dump valve and these are designed so as to provide for rapid filling of the dump tank. One obvious such design feature is a flap at the exit which lifts totally clear of the water causing no resistance to its flow. One of the most advanced dump tanks available is the **Fireflex Low Profile Tank™**.

Dunklin County, MO

The scene in April 2004 of a major fire involving waste tyres. Tyres have a high calorific value – about 40 MJkg^{-1} – and once ignited tend to burn smokily. At the Dunklin County fire other sorts of waste were involved in the fire at its early stages, including household refuse and paper. Strong winds at the time of the fire accelerated it, but no buildings were threatened. Tyre waste can be recycled to form "crumb rubber", which finds its way into products including Astroturf. In 2007 there was a major fire at a facility for such recycling in **Maricopa, AZ**.

Durrenasch, Switzerland

The scene in 1963 of an aircraft fire which caused 80 deaths. There was a tyre burst on a Caravelle aircraft taking off from Durrenasch airport, the pressure effects of which caused a fuel line to break. Fire resulted and the aircraft crashed killing all 80 occupants.

Düsseldorf Airport, fire at in 1996

Seventeen lives were lost as a result of a fire at the passenger terminal at Düsseldorf Airport in April 1996. The fire was started by hot debris from a welding operation. Amongst the factors believed to have contributed to the tragedy was absence of sprinklers from the part of the terminal where the fire began, and a paucity of smoke detectors across the terminal. Escape doors were electrically activated and could not be opened once power to the terminal was lost. There was a great deal of PVC wiring at the terminal, accounting for just under a quarter of the total mass of combustibles. Hydrogen chloride produced by this on conduction would have increased acridity of the smoke and (as Greenpeace have pointed out) caused some dioxin formation. (*See also* **Pyrodur®**.)

Duty of care, by fire services

This was invoked in an unusual case in Scotland where fire broke out in the same residential property within about 24 hours, each fire being attended by the local fire service. The owner of the property alleged that the fire had not been put out properly the first time. A court ruled that *if* that was so the owner could sue the fire service.

Dwarf dogwood

Botanical name *Cornus canadensis*. It occurs widely in British Columbia, but is used as a **firewise plant** as far south east as the Carolinas.

Dictionary of Energy and Fuels

Clifford Jones, Reader in
Engineering, University of
Aberdeen, UK and
Nigel Russell, Lecturer
in Chemical and Process
Engineering, University of
Sheffield, UK

This comprehensive new dictionary
comprises over 1300 definitions to
provide an extremely useful ready-
reference work on solid, liquid and
gaseous fuels, including information on the scenes of production of many fuels, such
as major coal reserves and large oil and gas fields.

The *Dictionary of Energy and Fuels* is a reliable reference work on fuel and energy
which will remain of great usefulness despite any future changes and trends in related
technologies.

Readership: This easy-to-use dictionary will be of great value to undergraduates,
postgraduates and professionals in fuel technology, industrial chemistry, chemical
and mechanical engineering and whenever a ready-reference on energy and fuels is
required.

> ...A staggeringly wide range of definitions is on offer. ...A list of some fifty
> references, a table of fuel-related units and conversion factors... round off the
> volume... a useful dictionary in a compact format which should satisfy those
> requiring a one-stop shop of information... The casual reader with an interest
> in the subject would also benefit... *Fuel*

ISBN 978-1904445-44-9   198 × 129 mm   368pp   softback   £16.99

E

E85, dispensers for

E85 is a blend of 85% ethanol in gasoline for use in spark ignition engines, having a very satisfactory octane number and currently finding wide use in the US and Australia, although it is not yet available in Britain. The Underwriters Laboratories are withholding certification of fuel dispensers for E85 until they have had the opportunity to assess the condition of such dispensers after significant periods of service. UL officials have performed such examinations not only in the US but in Brazil. Pending the UL response, individual states are issuing instructions for E85 dispensers.

Eagle 1928

A typical example of a **drum safety cabinet**, capable of holding one 55 gallon drum in horizontal orientation. The dimensions are 31.25 inch × 48.25 inch × 50 inch, giving a volume of 44 ft^3 ≡ 326 gallon, meaning that about a sixth of the space is actually occupied by the stored liquid. The Eagle range, a product of the Eagle Manufacturing Company in Wellsburg, West Virginia, extends in size to the Eagle 1955 cabinet which holds two 55 gallon cylinders in vertical orientation. Eagle products are available with a **fusible link** device fitted.

Eagle Point Refinery, West Deptford, NJ

The scene of an above-ground tank fire, believed to have been caused by lightning. The tank contained xylene in a quantity of 1.5 million gallons. The fire was treated with foam by fire fighters and there was abundant smoke.

Eagleroc®

Material for use in building, containing **gypsum** and manufactured by American Gypsum in Colorado. It finds particular application to modular building. In addition it can be used as a furnace liner.

East Orange, NJ

The scene in January 2008 of a fire at a service station in the very early morning. There were no injuries. Two dozen fire fighters attended the fire, believed to have begun at a car parked at the service station. Nearby residences were evacuated.

East Troy, WI

The Volunteer Rescue Squad in this town recently took delivery of a **rescue truck** based on the **Ford F-550** although the engine is not Ford, being in fact a Navistar unit of 325 hp. Of length 3.4 m, the vehicle in "non walk-in" having exterior compartments which carry fire fighter's needs.

EC225

Helicopter manufactured by the French-based Eurocopter, having recently become available from the manufacturer in fire fighting configuration. It can hold 1000 gallons of water in a suspended flexible tank. Other Eurocopter products have seen such use, including the EC 725, the largest in the range, a.k.a. the Super Puma, the Cougar and the Caracal. As the Super Puma, this helicopter is in fire fighting use in Hong Kong.

Ecstasy

Name of a cruise vessel on board which, in July 1998, there was a fire which began in the laundry room. The vessel was off Florida at the time and 54 people on board had to be treated for smoke inhalation. It was the second serious fire at sea to have begun in a laundry room in just over a year, and both were close to Florida! In April 1997 the vessel *Vistafjord*, having departed Florida for Portugal, had to divert to the Bahamas after an electrical fire began in the laundry room and one crew member died from smoke inhalation.

Eductor

Device widely used as a foam proportioner in fire fighting. Water for fire extinguishment is passed through an orifice and accelerated. The increase in kinetic energy is offset by a reduction in pressure energy. The part of the eductor upstream of the orifice has a connection to a container of foam concentrate, which, in response to a pressure difference, flows into the water and mixes with it, furnishing a foam. The water-to-foam proportions can be determined by the orifice size. An eductor has the advantage of not requiring electricity, all of the energy requirement being met by the water from the hydrant to which it is connected.

Effingham, IL

The scene in 1949 of a hospital fire which claimed 74 lives. Amongst the dead were babies delivered 24 hours earlier or less. Fibreboard in the building structure had been a major factor in the tragedy and thereafter manufacturers of fibreboard began to incorporate fire retardants into their products.

E-glass

"Electrical grade glass", so called because of its suitability, by reason of its fibre-forming capabilities and the high electrical resistance of fibres so formed, for electrical applications, especially with a heavy current load. The substance, which contains silica at 54% by weight and smaller amounts of calcium, magnesium and boron compounds, also finds application to the manufacture of fire-resistant materials including **PyroRope™**.

Egypt, train fire in 2002

The death toll was 383 when a cylinder of gas on board a train from Cairo to Luxor exploded in February 2002. The number of passengers was about double what it ought to have been.

Electrical surge protectors, fire hazards of

A surge (a.k.a. as a "spike") in the mains supply voltage will occur when some device bearing heavy current is switched on or off. The surge will be experienced by other devices receiving voltage from the same transformer. Where protection from surges is required a metal oxide varistor (MOV), primary constituent zinc oxide, is commonly used. This can change in composition with use over time, especially if one or more particularly severe surge has been received, and having changed in composition can increase in resistance and become an unsuspected heating element.

Electrochemical detection of carbon monoxide

This is conceptually straightforward. In a galvanic cell with sulphuric acid solution as the electrolyte the following processes occur at the respective electrodes:

$$0.5O_2 + 2H^+ + 2e \rightarrow H_2O$$

$$CO + H_2O \rightarrow CO_2 + 2H^+ + 2e$$

$$\overline{CO + 0.5\ O_2 \rightarrow CO_2}$$

Clearly a rise in the carbon monoxide concentration will be manifest as a change in the emf of the cell. Other substances which can be detected by this means include ethylene and ethanol vapour.

Electroluminscence

Phenomenon whereby visible light is emitted by a material in response to a current passing through it or to an electric field. An example is the light-emitting diode (LED) which of course finds very wide application to emergency lighting for evacuation in a fire and to the illumination of exit signs. Whereas with a conventional light source a filament must be heated to high temperatures, this is not so with electroluminescent devices, which therefore run on very low power. An exit sign so illuminated is likely to require of the order of 5 W, possibly less, of power. This of course is easily obtainable from a battery. In fact an LED releasing white light can put out 300 lumens per watt whereas a conventional, incandescent light bulb is likely to put out 10 to 20 lumens per watt (higher if it is set up as a **halogen lamp**).

Electroguard 30

Antistatic paint for use in covering walls and floors of areas where hydrocarbon vapour hazards exist, manufactured by Edson Electronics in the UK.

Elevated temperature flame limit (ETFL)

Some flammable mixtures will not at room temperature respond to an ignition source, which makes the assignment of a conventional lower flammability limit (LFL) impossible as these are determined at room temperature. Such a mixture might however respond to an ignition source if its own temperature is above ambient. The minimum concentration of the mixture in air which the mixture will ignite at a specified raised temperature is called the elevated temperature flame limit (ETFL) and this is clearly analogous to the LFL. ETFL values are sometimes determined for refrigerants, including **Refrigerant 403a**. The ETFL of a refrigerant might relate to the mixture as supplied or to a form of it having become depleted in the lower boiling constituents through evaporation losses as will tend to happen in sustained usage.

Elim, AK

The scene in 2005 of a fire on board a commuter aircraft carrying six passengers. The aircraft landed at Elim with one engine on fire. Passengers were evacuated safely. After extinguishment a close examination of the engines in which the fire had occurred was carried out. There had been replacement of a hydraulic pump 31 hours before the flight, and this had necessitated removal of a fuel pump for access and its subsequent replacement. Leakage of fuel from this pump caused the fire.

Elite

"Dedicated" fire fighting vessel (that is no towing or supply function), the largest such in the world. Built at a Hong Kong ship yard and entering service in Hong Kong harbour in 2002, it has eight **monitors** and three pumps and carries foam. In addition to such facilities for **external fire fighting** it has deluge for self-protection. It can safely take on board up to 100 persons additional to the crew, and these can of course be survivors of a vessel receiving help from the *Elite*. Note that **FiFi I**, **FiFi II** and **FiFi III** criteria are the same for "dedicated" fire boats as for vessels with fire fighting functions amongst others, e.g. an anchor handling, tug and supply (AHTS) vessel.

Elkland, PA

The scene in 2007 of a bakery fire believed to have started close to a fuse box. About 40 fire fighters attended. Smoke damage was considerable.

Elk skin, use of in fire fighting apparel

Fire fighting gloves described elsewhere in this volume have had cow hide or goat hide as one material: elk skin is also very suitable. An example is the FireMate™ Kangaroo/Elkskin Wrist Gloves which as well as elk skin uses kangaroo leather. These fire fighter gloves also contain **Crosstech®**, Kevlar® and **Nomex®**. Kangaroo leather features in several makes and types of fire fighter glove and is recognised as having good inherent strength and good moisture resistance.

Elliptical tank

An elliptical tank in a **water tender** has the advantage over other shapes of tank that it lowers the centre of gravity of the **water tender** and causes the centre of gravity to shift less with fluid movement within the tank. This makes for enhanced stability when the tender is being driven. A cross section of such a tank normal to its axis is an ellipse. (The tank is not *ellipsoidal* as it has flat or domed ends.) Such tanks can be fabricated from steel, from polypropylene or from **GRP** and capacities are up to 4000 litres. US fire truck manufacturers offering the elliptical tank configuration include **Alexis** (IL), Kovatch (PA) and US Tanker (WI).

Elmhurst College

The four dormitories belonging to this liberal arts college in northern Illinois have been fitted with sprinkler systems. The work was done over the long vacations of 2002 and 2003. The building into which sprinklers were installed ranged in age from 35 years to 80 years and differences in configuration and in building materials made for difficulties.

Emerald Airways flight 311

Difficulties with this flight from the Isle of Man to Liverpool began with leakage of flammable hydraulic oil into the cabin. The cabin crew moved 33 passengers to the back of the aircraft, affecting its balance. On approach to landing the nose wheel steering was found not to be working, and the aircraft returned to the Isle of Man where it made a safe landing.

Emergency Medical Technician (EMT)

Although most EMTs belong to ambulance crews, many fire departments in the US, including the FDNY, have EMTs amongst their members. Each state has its own training and certification and its own statement of the scope within which an EMT's duties fall. There is a scale of EMTs and ascent of the scale brings expansion of such scope. Taking New York as an example, there are four steps on the ladder from Certified First Responder to Advanced Emergency Medical Technician; at that level the job title is not EMT but paramedic. In any of the states of the US, EMT training involves supervised and monitored on-the-job experience and also a written exam. Some community colleges are accredited to teach Emergency Medicine to the level required locally for EMT licensing, otherwise such teaching will be at centres within hospitals. The FDNY pays EMTs (2008 rates) a starting salary of US$27 295, rising incrementally to US$39 179 for an EMT with five years of service. For a paramedic the ceiling salary is US$50 501. EMT is a lesser post than fire fighter but an EMT in the FDNY can, by means of a written exam, become eligible for reclassification as a fire fighter.

Emigrant Creek, MT

Scene in September 2001 of the crash of a **Vertol 107** helicopter in fire fighting service which killed all of the crew. The helicopter was one of 15 assigned to a fire in an area of forest where there had been prescribed burning by the **ping pong ball** method a few days earlier.

Encap®

Alternative to **Hypalon®** as an ingredient of a resistant cover for fire hose, a Niedner (now Tyco) product. A polyurethane resin, it has high resistance to abrasion and chemical attack from organic liquids. **Glotek™** is an example of a fire hose which uses Encap®.

Encore®

Medium-duty **rescue vehicle** manufactured by Pierce and available in many configurations. One recently supplied to the Great Valley, NY Fire Company has a 330 hp diesel engine and a power take-off (PTO) generation capacity of 350 kW. Like many such vehicles, it carries facilities for illuminating the fire scene.

Engine nacelle, fire precautions for

Broadly speaking, the nacelle is the "pod" by means of which a jet engine is attached to an aircraft wing. In the event of engine fire, the nacelle has an important role in protecting the aircraft and to that end is itself protected by extinguishant release should an engine fire occur. **Halons** are used for this purpose, although alternatives are being investigated in response to the Montreal Protocol. A nacelle will be protected from heat during an engine fire by a fire-resistant seal, and there is a move towards use of sealing materials which also display **intumescence**. An engine nacelle also needs to incorporate a firewall, a metal plate or covering which absorbs heat and enables it to be conducted way. Convective heat loss of any heat having found its way to the outside surface of the nacelle will of course be vigorous at aircraft speeds.

Envirograf®

Intumescent material produced in sheets for cutting to the size and shape required by a particular application. Such applications include metal door fittings. In application Envirograf® can, for example when placed between a door hinge and the door surface using a common set of screws, be compressed without loss of its intumescent activity. This gives Envirograf® and intumescent materials like it supplied in sheet form an advantage over those supplied in tubes for application like putty.

EnviroLift™

Hydraulic oil having the remarkably high flash point of 314°C. The base material is soybean oil and it has been developed for use in elevators.

Erickson S-64

Helicopter for fire fighting use, manufactured in Oregon, USA and able to carry in a suspended portable vessel 2650 gallons of water. Water can be drawn from a source as shallow as 18 inches and its subsequent release is controllable across a range. The S-64 has seen fire fighting service in countries including the US, Canada, Mexico, Borneo, Italy, Greece, France, Turkey and Australia.

Escape fire

Technique in forest fire or prairie fire fighting, involving creation by intentional burning of a vegetation-free refuge area for fire fighters. Unlike a fire break, it is not intended to halt the fire itself. That American Indians used the technique in grassland fires centuries ago is not in doubt. The higher thermal inertia of trees than blades of grass has made application to forest fires more difficult. Fire fighting leader Wagner Dodge used the method at the **Mann Gulch, MT** fire. There were criticisms of his having done so, some relatives of his dead colleagues even suggesting that the escape fire had led to their deaths. The consensus is, however, that the fact that Dodge himself survived is attributable to the escape fire.

Escape hood

Device for protecting a fire evacuee from inhalation of smoke. Commonly it will have a catalyst for carbon monoxide conversion to carbon dioxide and a filter for removal of particulates. Removal of other harmful substances including hydrogen cyanide is possible by means of an activated carbon. Often an escape hood is designed for single use only and will provide protection for up to about 20 minutes (as with the **S-Cap**). Some employers provide escape masks and increasingly frequently they are being made available in hotel rooms. They also have an important role in protecting the residents of homes for the elderly. Some escape hoods are designed to be suitable for inclusion in a traveller's hand luggage as a personal precaution (e.g. the **Evac-U8** which sells for about US$75).

Esco S-0700

Above-ground fire hydrant manufactured in Norway and having been in production, with modifications, for 50 years. It stands 1.1 m above ground with burial depth as required. It has twin outlets with 2.5 inch **Storz couplings** for connection to hose. These are accessible by opening a metal enclosure ("clam shell"), a variation on most designs where the outlets are exposed at all times. Kongsberg Esco AS are the sole manufacturer, although previously another manufacturer produced the device with a decorative exterior solely for Bergen.

Eska fire fighting gloves

These are made partly of **Crosstech®** as well as having Kevlar® lining and palms. Marketed internationally and in particular in the US, they are made in Austria.

Eucalpyptus trees, in North America

A contributory factor in the bush fires in Australia is the fact that trees of the native genus *Eucalyptus* burn so readily, partly because of the flammable oil released from glands in the leaves. Where eucalypts have been naturalised to North America, they are invariably seen as **highly flammable trees**. Species so introduced include *Eucalyptus cladocalyx* (sugar gum), *Eucalyptus globulus* (blue gum), *Eucalyptus rudis* (flooded gum) and *Eucalyptus viminalis* (manna gum).

Evac-U8

Escape hood, having the fairly typical performance characteristic of being able to protect for about 15 minutes. It is made of **Kapton**®.

Excel-FR™

Fire-retardant cotton fabric owing its fire retardancy to treatment with **THPOH**. It is used to make apparel for the metallurgical and petrochemical industries amongst others. Banox®, manufactured by Itex in Colorado, is a similar product.

Exolit

Family of phosphorus-containing flame retardants. Exolit AP and Exolit OP are composed respectively of ammonium polyphosphate and of certain antimony-organics whilst Exolit RP is composed of **red phosphorus**. Applications of Exolit products, manufactured by the multinational chemical producer Clariant, include polymers, resins, fibres and adhesives. Exolit AP contains **ammonium polyphosphate**.

External fire fighting

Term applied to marine vessels for fire fighting, being "external" to the scene of the fire, which might be another vessel (as in the case of the **Balmoral Sea**), an oil platform or an FPSO (Floating Production, Storage and Offloading) facility. An external fire fighting vessel might have that sole use or it might be a tug, coastguard vessel (as in the case of the **UT 512-OPV**) or offshore supply vessel with a supplementary fire fighting role. Performance criteria are volumetric water supply rate and "throw". Vessels are classified in terms of their equipment levels as **FiFi 0, FiFi I, FiFi II** or **FiFi III**.

A Pierce Contender pumper (courtesy Pierce Manufacturing)

F

Fairmount Sherpa

One of a family of very powerful tug vessels, entering service in 2005. Having been built by Niigata Shipbuilding and Repair in Japan, it was delivered to Fairmont Marine BV in the Netherlands. Fire fighting is at **FiFi I** level. Sister vessels, delivered in 2005–2007, are the *Fairmount Summit*, the *Fairmount Alpine*, the *Fairmount Glacier* and the *Fairmount Expedition*. Another product of Niigita Shipbuilding and Repair is the *Shin Chou Maru*, an anchor handling, tug and supply vessel (also **FiFi I**) owned by Fukuda salvage, which has accompanied two of the *Fairmount* vessels in taking an FPSO from Korea to Angolan waters.

Falkland Islands, fire service in

The year 1998 saw the centenary of a fire service in the Falkland Islands. In 1898 a volunteer brigade was formed in Port Stanley, and the brigade remained voluntary until after the Falklands War in 1982. It was, however, provided with successively more modern fire fighting plant. In 1985 the brigade was re-organised and now has a body of firemen a minority of whom are full time. The part-time firemen are paid a retaining fee and are expected to attend for weekly training sessions. The fire appliances acquired in 1985, demands on which have not been excessive, remain in use.

Fantraxx

Range of fans for use in **PPV**, available in 18 inch, 21 inch and 24 inch blade diameter versions. In each version the blade is made of polypropylene and rotation speed is around 3000 rpm. Honda 5.5 or 6 hp motors are used.

Far Sound

Anchor, tug handling and supply (AHTS) vessel entering service in 2007 as a "new build". It is classified **FiFi II** and registered to the Isle of Man. It is currently in use in the North Sea on a spot market basis. FiFi II is relatively rare for an AHTS vessel and **FiFi I** appears to be if not the norm the more common choice. Factors influencing the choice include the weight of the engines needed to drive the fire pumps and the space they occupy.

FE-13™

Extinguishing agent manufactured by DuPont, composed of fluoroform, CHF_3, which is of course the fluoro analogue of chloroform. It was developed as a substitute for **Halon 1301** and has negligible ozone depletion potential. Its action in extinguishment is due in part to the "neutralisation" of reactive intermediates including the hydrogen atom by the halogen component. FE-13™ and FE-227™ are also replacements for Halon 1301.

FE-36™

Extinguishing agent manufactured by DuPont, in a sense complementing **FE-13™** and FE-227™ in that it is primarily a replacement for **Halon 1211**. Its "aliases" are hydrofluorocarbon-236fa and HFC-236fa.

Federal Excess Property Program

Scheme by means of which equipment originally belonging to the Department of Defense can, via the USDA Forest Service, be made available on long-term loan to fire fighting groups in areas susceptible to forest fires. Such equipment includes **SCBA**, vehicles and pumps. Aircraft have also been loaned, for example to the California Department of Forestry. The scheme has been in operation for over 50 years. The only eligibility criterion for participation is that the fire department applying for equipment has responsibility for areas where vegetation ("wildland") fires occur: there is no maximum population requirement as there is with the **Volunteer Fire Assistance Program**.

Ferric oxide, use of in aerial fire fighting

Water from an **air tanker** is sometimes red in colour, so that its path to earth and (more importantly) where it comes to earth can be observed. The red colour is achieved by putting some ferric oxide in the water.

Ferrocene

Chemical compound of formula $Fe(C_5H_5)_2$, once promoted as an alternative to lead tetraethyl as an antiknock agent. It is used in fire protection as a smoke **suppressant**, for example in PVC. Organometallic chemists have synthesised many derivatives of ferrocene, some of which find application as smoke suppressants. One such has the IUPAC name β-ferrocenoylpropanoic acid.

FFFP

Acronym for *Film-Forming FluoroProtein*, which when mixed with water will create a foam for fire extinguishment. An example is Petroseal 3, which is suitable for hydrocarbon fires. The efficacy of a fire fighting foam for hydrocarbons depends on its miscibility with the hydrocarbon itself and there can be difficulties if the hydrocarbon is substituted or partially oxidised, as with an alcohol. It is for such cases that Alcohol Resistant Film-Forming FluoroProteins (AR-FFFP) have been developed; an example is **Niagara 3-3** as is one of the **Foamousse** group of products. The respective viscosities are also a factor in matching a foam to a flammable liquid. (See also **Foam/water sprinkler**.)

Fibre glass ladder

A device for preventing **fire ground electrocution**. Such a ladder will not be composed entirely of fibre glass but will have side rails made from fibre glass. The metal rungs will be electrically isolated from each other and from any metal attachments at the top of the ladder. If such an attachment contacts a power cable, there will be no conduction along the ladder, therefore no threat to a fire fighter on it.

FiFi 0

A.k.a. FiFi½, the least stringent of the four classifications of level of equipment on a fire fighting vessel and the baseline in that a vessel of FiFi 0 is not in the strict legal sense "classified": the term simply means that the vessel is either built or adapted for fire fighting in addition to other duties, e.g. those of a tug, which it might fulfil, as with the vessel **Barbados II**. Usually a vessel designated FiFi 0 will have a minimum water application capacity of 300 m^3 per hour, possibly exceeding this by up to half an order of magnitude. Such a vessel may or may not have a foam system.

FiFi I

This classification requires two devices for water application ("**monitors**") each with a throw distance of at least 120 m and a maximum height of 45 m. These must be remotely controlled, that is controlled from a site on the vessel remote from the engine room. A tug might be equipped to this level as might a vessel solely for fire fighting.

FiFi II

This also requires two water application devices, with throw and maximum height respectively 180 m and 110 m. The water supply rate must be 7200 m^3 per hour or greater. A vessel of FiFi II designation will often be a tug or an offshore supply vessel with a dual role, as with the *Geonisio Barroso*.

FiFi III

Vessels sufficiently equipped for this classification are not numerous, an example being the **Maersk M Type**. To qualify for FiFi III a vessel has to have at least three water application devices, the throw and height criteria being the same as for **FiFi II**, the volumetric flow requirement being 9600 m^3 per hour.

FiFi, dual classifications of

In addition to **FiFi I**, **FiFi II** and **FiFi III** the classifications FiFi I and II, and FiFi I and III are recognised. FiF I and II differs from FiFi II in having a lower degree of self-protection than FiFi II but the same external fire fighting capability. The distinction between FiFi I and III and FiFi III is on the same basis. **Active Lord** and **Bonassola** are examples of FiFi I/II vessels.

Fiona

Hydrocarbon-bearing vessel upon which there was a major explosion in August 1988. The ignition source was steam, which created static electricity, which was able to build up on a temperature sensor lacking earthing. A recommendation from the follow-up was that such sensors should have a label declaring the need to earth them.

Fire appliances, power units for

In the US, major suppliers of diesel engines for fire trucks include Detroit Diesel, Cummins and the International Truck and Engine Corporation ("International"). An aerial requires an engine of about 500 hp (37 kW), and the **SPH 100**, manufactured by **Sutphen**, might have a Detroit Diesel 515 hp unit or a Cummins 500 hp unit. The aerial manufactured by **Seagrave** recently delivered to **Redwood, CA** has a Detroit Diesel unit of 500 hp. As an example of the power requirement of a pumper, the **Contender** will commonly have an engine of 350 hp. Customisation and adaptation to particular needs of users precludes generalisations about engine size however, and one recently delivered Contender has a 500 hp engine manufactured by International. In the UK and some other parts of Europe **Scania** engines are installed into Scania fire trucks, consistently with the principle of integral design. The Dennis **Sabre** uses a 295 hp unit made by Cummins Euro and the **Dagger** also uses a Cummins engine. The **Titan** "four by four" **ARFF** vehicle has a Detroit Diesel 665 hp unit. ARFF vehicles manufactured by the **Morita Corporation** use engines in the range 540 to 1024 hp, the lower figure being for a 24 tonne ARFF and the higher for a 44 tonne ARFF. Some versions of the **Striker** ARFF vehicle have 950 hp and the top-of-the-range **Colet Jaguar** has 1600 hp. At the other extreme the **fire**

bluster, which weighs about 10 tonnes, has an engine not exceeding 175 hp. (*See also* **American LaFrance, recent deliveries by; Fire appliances, used and rebuilt**.)

Fire appliances, used and rebuilt

Fire trucks in service with city fire departments and the like do not sustain high mileages and are usually garaged indoors. They receive maintenance at much more frequent intervals than most other vehicles do. The life expectancy of a fire truck is therefore long unless (as very often happens) obsolescence necessitates its replacement, in which case it might be saleable, either as is or after refurbishment and re-equipping. It has been noted in another entry that volunteer fire authorities, for obvious financial reasons, tend to buy used appliances and in rural areas of the US acquisition of used fire trucks is common as at **Sikeston, MO**. Manufacturers including Pierce offer pre-used fire trucks for sale and **Seagrave** offer users of their trucks refurbishment of existing trucks as an alternative to purchase of new ones. The **Public Protection Classification (PPC™)** system of rating a locality for fire insurance purposes does not specify any maximum age of fire appliance. The table below gives details of a representative selection of such having recently been sold, or made available for sale, by Brindlee Mountain Fire Apparatus in Alabama,[13] a leader in the used fire truck business who at any one time have a wide range of stock varying in asking price from 10 thousand dollars or less to over a quarter of a million dollars.

Make and type of appliance and year of initial use	Details and comments
E-One aerial, 2002	Three-axle aerial, 105 feet ladder reach, with pump and water storage facilities. Easy to set up as a **quint**
American LaFrance pumper, 2002	Cummins 330 hp engine. 1500 gpm Hale pump. 750 gallon water tank. Previously in use in Schenectady, NY
Saulsbury (now part of E-One) pumper, 2002	Cummins 330 hp engine. 1500 gpm Hale pump. 500 gallon water tank. Asking price US$249 000
American LaFrance **attack apparatus**, 2000	**Ford F-550** base
Spartan **quint**, 1993	Two-axle **quint**, 75 feet ladder reach

13   http://www.bmfallc.com/index3.htm

Pierce **rescue vehicle**, 1992	Walk-in type, 9.5 m in length with Cummins 350 hp engine
Sutphen **quint**, 1991	Two-axle quint, 75 feet ladder reach
E-one pumper, 1989	Detroit diesel 445 hp engine. 1500 gpm Hale pump. 500 gallon water tank. Asking price US$59 000
E-one pumper, 1978	General Motors base, gasoline engine. 250 gpm Hale pump. 500 gallon water tank. Asking price US$12 000
FMC* pumper, 1978	Gasoline engine. 750 g.p.m. Hale pump. 500 gallon water tank. Asking price $5000

* Ceased production of fire trucks 1990.

Other major US suppliers of used fire appliances include Fenton Fire Inc. in Michigan whose current stock includes one of the few remaining trucks built by **Pirsch**. The fire appliance scene in the UK is very different from that in the US: different types of appliance (one would not for example expect to find a **tractor-drawn aerial** anywhere in the UK) and different organisation of fire fighting authorities and groups. An appliance having completed its term with a fire brigade in Britain is seldom saleable within that country (except perhaps to a refinery) and a common fate for such an appliance is for it to be donated for use in a poorer country. Countries having received fire appliances in this way include Poland, Romania and Peru as well as several African nations.

"Fire-attack mode"

Modus operandi whereby **attack apparatus** and the **water tender** are one and the same fire truck, which must therefore be self-sufficient in all respects, carrying hose, ladders and possibly foam concentrate.

Firebird

Trade name for fire extinguisher cabinets manufactured by Jo Bird, England. They are made from **GRP** for the offshore oil and gas industry, where weather conditions are often severe. A single cabinet can be supplied to hold one extinguisher or two. Fire hose cabinet and fire fighters' clothing cabinets made of GRP are also sold under the Firebird name, again for offshore use. Each product range is of course an example of **detention fire equipment**. Jo Bird has recently landed contracts for supply of such equipment to offshore oil and gas facilities in Thai, Australian, Angolan and UK waters. The company has also supplied operators at Sakhalin. The Firebird range is complemented by the **Toughstore** range.

Fire Blockade™

Newly developed fire extinguishing agent developed and manufactured by Chemguard in Mansfield, TX. It contains a protein as well as wetting agents and is combined with water at about 1% concentration. It is suitable for wide application including forest fires.

Fire bluster

Range of fire trucks made by **Zil**, currently in production. Each has a tank and a pump but lacks aerial ladders. Water release at 600 to 700 g.p.m. is possible using a pump manufactured by the Lithuanian company Iskada.

Fire Boss

Newly developed (by Wipaire, St. Paul, MN) fire fighting aircraft having made its US debut in April 2007, although several were in service in Europe and in Canada by that time. It has floats enabling it to land on water and thereby (like the **Hawaii Mars**) take on water without returning to base. Having a water capacity of 800 gallons it can "take over" whilst larger **air tankers** return to land for refilling with water. It can be fitted with foam proportioning capability, as is the case with the two recently entering service in Sardinia. The acronym SEAT – single-engine attack tanker – has been applied to the Fire Boss.

Fire collar

Device for preventing flame propagation along a plastic pipe. It comprises a steel cylinder containing material capable of **intumesence**. When a flame encounters it, the intumescent substance swells and crushes the plastic pipe. Commonly a flame collar will be fitted where a plastic pipe goes from one floor of a building to another. If the pipe ignites on one of the floors, the fire collar will prevent spread by this means.

Fire damage, financial losses due to in UK and US

Expressed as a percentage of the Gross Domestic Product, this is around 0.13% (1.3 thousandths) for the UK. The fire services in the UK cost about 0.21% so the estimated total annual cost of fires is 0.34% or US$8150 million (≡ £4100 million). In the US the total annual cost estimated is between US$45 000 million and US$50 000 million. (See also **Fire services, costs of in selected countries**.)

Fire damper

Device which can be installed into a duct at a point where it goes through a wall or floor to prevent the fire from spreading via the duct. Under normal circumstances the damper is open and allowing the duct to fulfil its role in ventilation, process gas discharge or whatever. If a preset temperature, say 70°C, is measured by a temperature sensor at the duct, the damper will close. Having done so it will block the duct to flame propagation, though not of course for an indefinite period. A particular damper will have a period for which, having been tested to a standard, it can so act. A damper in its simplest form is just piece of steel, possibly incorporating slats. An alternative is for the duct to contain a plate across one cross section with orifices in it allowing gas movement under normal conditions. If this exceeds a certain temperature, an intumescent material mounted on the plate comes into action and blocks off the orifices. Again, configuration as slats is possible. **Astro Damper** is an example of a fire damper which works by **intumescence**. An advantage of the intumescent type is that they have no reliance on electrical power supply and so will not fail if that supply is not functioning at the time when closure is required. Of course, a non-intumescent fire damper will not be affected if power fails after closure.

Firefighter® 5 Series

Engine driven pumps for fire fighting, available with 6, 9 and 13 hp engines and light enough to be portable. The middle-of-the-range model delivers 95 gpm at 80 psi. They are of Australian design and manufacture.

Fire fighter hood

Protective piece of apparel covering a fire fighter's head and extending over the shoulders, and also below the collar as a "bib". This and the other items of protection are collectively called the fire fighter ensemble (FFE). **Nomex®** is widely used in fabrication of such hoods as is Kevlar®: both are **aramid fibre** products as is **Carbon X®**. **PBI™** is also used in hoods.

Fire fighters, grave markers

Eagle Flag, a business based in Weymouth, MA, provides amongst other things grave markers for US citizens having lost their lives whilst on military service, there being different designs of marker for the First World War, the Second World War, Korea, Vietnam and so on. Eagle Flag also supply markers for the graves of fire fighters killed whilst on duty. A number of designs are available one of which, made of bronze, has an image of a ladder on one side and of a hydrant on the other. This sells for about $US40.

Fire fighters, mortality rate of in US

The National Fire Protection Association, from their base near Boston, record that in 2005 87 US fire fighters died whilst on duty. Of those 87 only 12 occurred close to a fire and might possibly be attributed directly to the fire itself. Of the others, 47 were due to "cardiac events" and not all of these were during an alarm; some occurred between alarms when the fire fighters were "on duty" at base. A helicopter crash accounted for a number of deaths among US fire fighters in that year. The 2007 figure was raised by the deaths in a fire in Charleston, NC of nine fire fighters in June of that year. The fire was at a furniture warehouse the roof of which collapsed into the area where the fire fighters were working.

Fire fighters, protection of from traffic

When a vehicle fire occurs on a multi-lane highway, its attendance by the fire department will not necessarily necessitate closure of all lanes. Even so a decision is sometimes taken to close all lanes so that fire fighters are not in danger from speeding traffic. The right so to close a major highway in one direction and divert traffic from it was invoked in Tampa, FL in March 2006 when fire fighters were called to a fire in a motor caravan.

Fire fighters, tertiary qualifications for

In the US, advancement above about mid-level on the fire fighter scale requires formal qualifications on top of in-house training, ideally a degree at Baccalaureate level. The United States Fire Administration (USFA), a Federal Government body, encourages fire fighters to obtain such quali-fications and accredits certain courses. One such course is the Bachelor of Science in Fire Service Management at Southern Illinois University in Carbondale, IL. Similarly Eastern Oregon University, in partnership with two nearby academic institutions, offers a Baccalaureate degree in Fire Services Administration, with courses in areas including personnel management, fiscal management, IT and local government. Arizona State University offers a Bachelor of Applied Science degree in Fire Service Management directed at part-time participants currently serving as fire fighters. Cogswell College in California offers two degree programs in Fire Administration and Fire Prevention/Technology. Each can be done on a distance learning basis. There are several other such courses across the US accredited by the USFA, many of which can be done wholly or partly on-line. Documented professional development by way of internal training or NFPA programs can count towards the requirements for admission. The degree course at the Southern Illinois University in Carbondale was successfully completed by David B. Foreman, who since 2006 has been State Fire Marshall for

Illinois. He is also an **Emergency Medical Technician (EMT)**. Several other state fire marshals, including those for California and Texas, have Bachelor's degrees.

Fire fighting front, communication between fire fighters at

For fire fighters to be able to communicate with each other whilst wearing breathing apparatus is an obvious advantage. Accordingly an amplifier is often fitted to a **SCBA**. This is battery operated and a light indicates "low battery". This is mounted on a "heads-up display" (HUD) so as always to be visible to the fire fighter. This display also contains a light which comes on when time for renewal of the air cylinder is approaching. Many fire departments use SCBA so equipped, and some purchasers of SBCA, including the US Navy, make it a positive requirement. Sometimes an SCBA is fitted with a warbling alarm by means of which a fire fighter can inform others that he is in difficulty.

Fire fighting hose, materials for

Commonly, fire hose carried by fire trucks for direction of water at a fire consists of an inner layer of rubber with a woven outer structure, meaning that the basic design has not changed fundamentally since the patent for a fire hose granted in 1821 to **James Boyd**. The reason for a woven outside layer is that this enables the hose to be stored flat thus saving valuable space on a fire truck. The diameter when occupied by water is typically 2.5 inches (64 mm). Cotton served for very many years as the outer layer of fire hoses but such materials as polyesters and nylon have replaced it: **aramid fibre** has also been used for this purpose. Similarly, synthetic rubbers, including several of those based on butadiene, have replaced natural rubber as the lining material. Silicone rubber has also been so applied, in the case of **Firestream®, EPDM**. Depending on the pressure to be withstood, the outer woven layer may be single ("single jacket hose") or double ("double jacket hose"). Hose having the composite structure described is not restricted to mobile fire units[14] but is often present at stationary fire fighting installations at places including chemical plants, airports and harbours. NFPA or other standards apply to such hose as well as to **hard suction hose**. Hose composed of a single material is **unlined hose**. **Nitrile** finds considerable application to fire hose manufacture.

Fireflex Low Profile Tank™

Dump tank, from the manufacturer of the **Bambi bucket®**. For use in fire fighting from the ground, it has a solid float collar which rises as water is

---

14   These are sometimes termed WOW (= water on wheels).

admitted, eliminating the need for a frame. By means of this it "assembles" naturally and this saves time.

Fire ground electrocution

Term applied when a metal fire fighting ladder contacts a live power cable, killing a fire fighter in contact with the ladder. There have been many occurrences, including one in New York City when three fire fighters carrying a ladder vertically slipped on ice and the ladder contacted a cable spanning the street, causing the death of one of the fire fighters. Three fire fighters were killed by fire ground electrocution in a different part of the US in 1998. In fact they were not attending a fire: the ladders, made of aluminium, had been made available as a community service to help with the painting of a local church building. Power *within* a building is often turned off during fire fighting. In some parts of the US the electrical supplier is sent for and a trained representative turns off the power as required, possibly in stages. Electrical risks are eliminated where a **fibre glass ladder** is used.

Fire hose, life expectancy of

From the entries in this volume which deal with the various sorts of fire hose in current use, it is clear not only that such hose has evolved over the decades to very high standards of construction and reliability but also that R&D into hose is ongoing. It has been suggested – there is a long way to go before the suggestion becomes a formal recommendation – by one of the committees of NFPA that *all* fire hose in the US should be taken out of service after ten years of use. There is much further still to go before a formal recommendation, if and when made, becomes the subject of legislation. Many fire departments in the US are in fact using fire hose older than ten years, especially in rural areas. It is clear from such reports as there have been of the matter that the issue is primarily one of risk assessment having regard to the fact that no piece of fire hose is totally reliable even when brand new. (It is helpful to note that one type of fire hose – the **Darley 800** hose – comes with a ten year warranty from the manufacturer.) Fire hose finally taken out of service can be used to make **chafing pads**.

Fire hose lining, condition monitoring of

One of the difficulties with hose is separation of liner and jacket, which is why a hose not using an adhesive such as **Viking fire hose** or **Carry-Lite** is in some ways preferable. For hose comprising a single woven jacket with a liner and adhesive in between, there is a test which can be applied to determine the state of adhesion. A 1.5 inch strip of the lining is separated from the jacket and a 12 pound weight applied. The hose is deemed to be in serviceable condition if the rate of the resulting bifurcation of jacket and liner does not exceed 1 inch per minute. (It is by no means unknown for a piece of liner, having become detached during use, to travel to the nozzle and block it.)

Fire hose reels, manual and automatic

When an automatic fire hose is mounted on a reel supported on the wall of a building, the act of unreeling the hose will open a valve and the reel is said to be automatic. Otherwise the valve has to be opened by the operator and the reel is classified as "manual". Often a particular reel will be made available by its manufacturer in automatic or in manual form, as with the **Macron 55 Series**.

Fire hydrant, reliability of

The reliability of a hydrant will never be 1.0 (or in percentage terms 100), nor is an actual reliability such as 0.95 a hard number. We intuitively expect that it will drop if, for example, the interval between routine checks is increased. Maintenance procedures enable the reliability of a hydrant to be kept as high as is consistent with the "reasonably practicable" concept in risk analysis. Failures do occur. For example in Northdale, FL in August 2007 fire fighters attending a house fire attempted to draw water from a hydrant but to no avail and instead had to use one at a nearby football field. The hydrant which failed was about 30 years old, not an excessive age. It had undergone a routine check about two months earlier and no malfunctions had been reported. After examination of the hydrant, the difficulty was attributed to insufficient lubrication.

Fire hydrant thawing unit

In spite of precautions, including the use of dry barrel hydrants, fire fighters do sometimes find that their work is slowed down by ice and frost in a hydrant and possibly in hose. A fire hydrant thawing unit can overcome this by directing steam at the hydrant. Such a device was of enormous help to fire fighters attending an early morning fire in Brooklyn, New York in February 2007. Each borough within New York City has one fire hydrant thawing unit. There is also such a unit in Boston.

Fire insurance, first in the UK

This was a spin-off from the Great Fire of London in 1666. Insurance companies then understood that it was in their interests to provide fire fighting teams, and a particular building would have a badge attached declaring with which company it was insured.[15] Fire teams from one insur-

---

15 Examples can be seen on: http://www.glosfire.gov.uk/sections/schools/school_rc_insur-ance.html.

ance company would not in general attempt to extinguish a fire in a building insured by another. The fire fighters were part-time, their first job being on boats on the Thames.

Fire main

The fire main of a ship comprises a pump to draw in sea water for extinguishment purposes, pipes (commonly plastic) for its distribution and connections for hose. **Unlined hose** is often used in ships.

Firemaster® 552

Trade name of a flame retardant developed for polyurethane foams, though it is also used for furnishings and car interiors. It has phosphorus and bromine amongst its constituents. An issue with fire retardants for polymers is that they can cause impairment by discolouration. One selling point of Firemaster®552 is that when used with polyurethane foams it causes such discolouration in a very low degree, if at all. Firemaster®552 was developed and is produced by the Great Lakes Chemical Corporation,[16] which has its business HQ in Lafayette, IN. Another Great Lakes product is Firemaster® 2100 which finds wider application than its namesake, being used in polymers including polypropylene and polybutylene terephthalate as well as in adhesives and coatings. Firemaster® T33P, like **Antiblaze®** 195, is used in car upholstery and the two are directed at the same markets. Firemaster® BP411 is a **brominated polystyrene**.

Fire pillow

This is used to protect cables passing through ducts and across trays. A pillow will be made of an insulating material such as fibre glass together with an intumescent substance, the action of which will inhibit propagation and smoke movement in the event of fire. One example amongst many is the TREMstop PS, which has the advantage of freedom from dust production on operation, making it suitable for use in computer rooms and other locations where atmospheric particulates could cause non-thermal damage.

Fire services, costs of in selected countries

The Gross Domestic Product (GDP) of a country for a particular year is the money spent within it on goods and services. The cost of the fire services can therefore helpfully be expressed as a percentage of the GDP and in the UK it is about 0.2%, that is two thousandths of the GDP. In the US and Canada it is about 70% higher, yet the death rates from fire in these coun-

---

16   Great Lakes is the world's largest producer of flame retardants.

tries expressed on a unit population basis are higher than that for the UK. Figures for seven selected countries are in the table below.

Country	GDP/US$ millions[*]	Approximate expenditure on fire services/US$ millions
USA	13 194 700	33 000
Canada	1 275 273	4500
Japan	4 366 459	15 000
UK	2 398 946	5000
New Zealand	104 607	150
Netherlands	670 929	1000
Denmark	276 400	200

* World Bank 2006 figure.

If the figure for the USA in the right hand column is divided by the population of the USA, it transpires that every member of the population – man, woman and child – pays US$100 per year for the benefits of the fire service. (*See also* **Fire damage, financial losses due to in UK and US**; **Volunteer fire fighters, financial value of the work of.**)

Fire Sprinkler Tax Incentive Act

Senator John D. Rockefeller IV Democrat, West Virginia was one of the sponsors of this Act. Its aim is to make sprinkler systems fitted to buildings fully depreciable over a five-year period. This will enable part of the cost of sprinkler installation to be recoverable on a realistic time scale. The fact that the **Station Nightclub** was *not* fitted with sprinklers was raised as a point in favour of passing the Act.

Firestream®

New fire hose product, having a single jacket made of polyester and a lining made of the synthetic rubber EPDM (ethylene/propylene/diene monomer). The outside is treated with Encap™, which as well as protecting from chemical damage and water absorption provides abrasion resistance.

Fire truck, appliance, engine

The first two of these expressions are approximately synonymous and are generic. The third strictly applies only if an engine for pumping water (such as one from the **Godiva World Series**, or one manufactured by **Waterous**) is fitted, in which case the word "pumper" might be used. The broad term apparatus is also encountered. A fire appliance with aerial ladders might be referred to as an "aerial" or as a "turntable ladder" or simply as a TL or a TTL. Local customs prevail in such nomenclature and have to be observed. A particular term relating to fire apparatus can however mean different things to different people even within the same country! (*See also* **Hydraulic ladder; Quad; Quint; Rescue vehicle; Tractor-drawn aerial; Turret; Water tender; Water ladder; Water tower truck**.)

Fire truck, conversion of from water to Compressed Air Foam

The merits of **Compressed Air Foam** as an alternative to water are such that the conversion of existing fire trucks from one to the other has in fact taken place in a number of US localities. A "conventional" fire truck needs to be able to pump water. One for **CAF** needs a "proportioner" whereby it can produce the required foam by blending water and foam concentrate in the correct proportions. It also needs an air compressor or alternatively a means of storing air under pressure. Some fire trucks as old as ten years have been economically adapted from water to CAF. Some fire departments have a policy of moving gradually from water to CAF by attrition, that is when a water fire truck is taken out of service it is replaced with a CAF one and the CAF capability of the department rises accordingly.

Firewall® ARFF

Fire fighter gloves. Features include a wristlet made of **Nomex®**. Other constituents include Kevlar® and Crosstech®.

Firewise plants

In such places as the Rocky Mountains, homes at the "wildfire/urban interface" need special protection from fire. One measure is the choice of plants for the garden which have a good fire resistance and in the US such plants are said to be "firewise". Two broad principles relating to fire resistance are that a high moisture content is favourable as is sparse rather than luxuriant foliage. Many plants have been tested and evaluated for fire resistance. Some of these have their own entries in this volume, e.g. **chokecherry**, **crested wheatgrass**, **heavenly bamboo**, **mountain mahogany** and many others. Firewise plants will provide protection only if suitably landscaped, such factors as the distance of the plants from the building being protected and the degree of slope being relevant. Frequent watering might be required to maintain the high moisture contents.

Firex® 5700

Heat detector which uses a thermistor and requiring mains electricity supply, although a 9 V battery provides for continued functioning in the event of mains failure. Like other heat detectors, it is recommended for use in environments in which a smoke detector might be needlessly activated by dust or fumes.

Fixed gallonage nozzle

This term for a nozzle used with fire hose is both self-explanatory and yet ambiguous. The delivery at a nozzle depends upon the water pressure, so to declare a nozzle to be "fixed gallonage" under all circumstances of flow is incorrect. A truly "fixed gallonage" nozzle is characterised by a gallon per minute (g.p.m.)–pressure data pair, e.g. 250 g.p.m. at 50 psi. Such a nozzle comes into operation only when the pressure upstream of an orifice ("flow disc") is the set pressure so that the corresponding g.p.m. will apply. Whilst this precludes a change in g.p.m. during operation, a nozzle can be supplied with several flow discs to vary the set pressure and hence the g.p.m. The above data pair is for a particular fixed gallonage nozzle of the "Metro II" type manufactured by TFT (Task Force Tips), Indiana. This is supplied with a set of flow discs providing a g.p.m. range from 125 to 325. TFT also make attachments whereby their nozzles can operate as **aspirated foam nozzles**.

Flacavon®

Diverse range of flame-retardant products from the German concern Schill and Seilacher. It includes Flacavon®H, which is an antimony-containing retardant for leathers. Flakavon® FK, also antimony containing, is for used on cotton. Flacavon® R is a single chemical compound of formula $C_9H_{15}Br_6O_4P$ having counterparts in the **Fyrol**™ range amongst others. Flacavon® AZ, having an organophosphorus reagent, has possible interchangeability with the product **Antiblaze®** CU.

FlameGard®

Flame retardant and **smoke suppressant** composed of alumina trihydrate (ATH) at particle sizes in the range 22–26 μm. It is manufactured by Almatis, whose HQ is in Frankfurt and finds application *inter alia* to carpets. Let it be noted that the particle sizes are fairly large: ATH is available in particle sizes of the order of 1 μm or even lower, as with Spacerite®, also manufactured by Almatis. This is incorporated into paints which use a titanium oxide pigment, and contains ATH particles of average size 0.25 μm.

FlameOut™

Extinguishment device for kitchen fires comprising a cloth having been treated with a proprietary flame-retardant substance. It has been the subject of independent assessment by ASTM methods, and the findings posted on the Internet.[17]

---

17 Go to: http://flamebustersofkansas.com/Reports/warnock-fabric.html

"Flame retardant" and "fire retardant" – synonymous?

From general experience over the years and having consulted technical dictionaries and a thesaurus, the author believes the two terms when used as nouns to be synonymous. The following relevant points can however be made. These substances usually prevent incipient, not developed, combustion, and "fire retardant" might be preferable for that reason. There is however no reason to shun the term "flame retardant" as it is very widely used. In scientific and engineering vocabulary meaning is derived from established usage rather than from strict application of terminology or etymology. The term "flame retardant" would describe a substance which causes a flame, once propagating, to do so more slowly than it would have done in the absence of the retardant. So seldom is the term "flame retardant" used in that sense that the rule whereby the terms "fire retardant" and "flame retardant" mean the same is proved by the exception! The adjectival term "fire-retardant" is applied to materials which have been treated so as to make them fire resistant, e.g. cotton treated with **THPOH**.

Flame retardants, equivalence and interchangeability of

Many chemical companies are in the flame-retardant business and therefore have to compete with each other. To do so effectively a manufacturer has to indicate which of its products is closest in composition and performance to a retardant, manufactured by a rival manufacturer, which a potential customer is currently using. Consequently tables of equivalence of retardants have developed (and can be found on the web sites of the respective manufacturers), and we note that the US manufacturer ExpoMix declares its product PUMA® 4020, composed of tris(1,3-dichloroisopropyl) phosphate, to be "similar" to **Antiblaze**®195 and its product PUMA® 4030, tris(2-chloroethyl) phosphate, to be "similar" to **Antiblaze**®V6. "Similar to" means "at first consideration interchangeable with", there being discussions and possible trial required to confirm such interchangeability. If the two rival products being considered have conformed, for particular applications, to recognised standards issued for example by NFPA, that supports the view that they are equivalent. ExpoMix have also established "similarity" between certain of their products and those of the **Fyrol**™ range and the **Flacavon**® range also features in cross-referencing of different retardants.

Flame Spread Index

This is defined in ASTM Standard E 176 as well as in NFPA and UL Standards. Two benchmark materials are used: fibre-reinforced cement board with an index of 0 and Red Oak Flooring, with an index of 100. The test apparatus is adjusted, in terms of air flow, to burn the whole length of the Red Oak reference sample in 5.5 minutes. The subject sample is then placed in the test equipment under the same conditions of air flow and the extent of flame spread over a 10 minute period divided by the (standardised: the thickness is also specified) length of the Red Oak sample and this quotient is multiplied by 100. So for example, a Flame Spread Index of 40 is interpreted as explained in the box below.

Flame Spread Index/100 =
distance of fire spread in 10 minutes/length of the Red Oak
reference sample

If the Red Oak Sample has length 24 feet (7.3 m), the fire in the
subject material has spread 0.4 × 7.3 m = 2.9 m, in 10 minutes.

Flame Spread Index is obtained from test results by the reverse of the calculation above and is always rounded to the closest multiple of 5. The value can of course exceed 100: it rarely in materials of interest exceeds 200.

Flame Stop I™

One of a group of proprietary fire retardants developed for cellulosic materials and manufactured by Flame Stop Inc. in Fort Worth, TX. The retardant, which is water based, penetrates a cellulosic substance and, should that substance ignite, prevents pyrolysis thereby depriving the combustion of the ignitable vapour-phase materials, which at least initially are the primary fuel. Flame Stop I™ is water soluble and can be applied like paint to a wooden surface in which case it affords protection for a period of years without renewal. If applied to a cellulosic fabric such as cotton in the form of curtains or carpets, it will withstand dry cleaning without loss but not laundering in water. The product finds application to wooden fitments in homes, shops, hotels and restaurants.

Other products in the range include Flame Stop II™ which was developed for outdoor use, having an ingredient which binds to a cellulosic structure making it resistant to diffusive loss on water penetration. Flame Stop III™ differs from the others in that it is intended to be applied not alone but as a blend with paint. Flame Stop I IM™, unlike the others, works by **intumescence**.

Flamex PF

Spray-on flame retardant for fabrics including curtains, carpets, blankets and upholstery. It is also suitable for wood. Manufactured by National Fireproofing Inc. in Coal City, IL, it contains phosphorus compounds as the retarding agent.

Flashover, criterion for

It was once a widely accepted rule of thumb that flashover – transition of a fire from being localised to involving the whole enclosure, with accompanying large increase in heat-release rate – occurred at a threshold heat-release rate of 1 MW. This is reasonable for calculation purposes where no more precise value is known, but the fire research community and journal editors have tended in recent years to discourage the close identification of flashover with attainment of 1 MW. There have been fires where from computational fluid dynamics it has been concluded that the heat-release rate at flashover exceeded this by half an order of magnitude. For risk analysis purposes a value of 1 MW is likely to err, if at all, on the low side, thereby predicting a flashover hazard where there is not one, which is better than vice versa.

Flexflat

Trade name of a type of fire hose, made of PVC reinforced with polyester, analogous to hoses made of **textile-reinforced rubber**. It comes in diameters up to 2 cm. The pressures at which Flexflat can be used are fairly low, up to about 75 psi, depending on the diameter. This is comparable to pressures at which **Vinylflow** can be used. This is in fact made of PVC but, unlike Flexflat, has a woven outer layer. Either hose is almost an order of magnitude below **Pro-Lite** or **Caspian 2026** in pressure of operation. Flexflat shares with hose materials including **Storex Laylite** the advantage that absence of a woven layer eliminates the need for drying after washing.

Flow proportioners for foam, mechanical power for

A foam solution for fire fighting is often made at the scene of a fire using water from a hydrant and foam concentrate solution carried. This requires use of a proportioner of which there are a number of designs and configurations including the **pressurised bladder**. Such proportioners are available which do not require electricity but obtain mechanical power from the pressure of the water exiting the hydrant. The calculation below gives an indication of the order of magnitude of mechanical power obtainable by this means.

From the steady flow equation assigning all of the energy to "pressure energy", rate of supply of energy = $(P_2 - P_1)/\sigma$ Jkg^{-1}

where P_2 = water pressure at hydrant exit (Nm^{-2})
P_1 = water pressure at proportioner exit (Nm^{-2})
σ = density of water (kg m^{-3})

Use the arbitrary but reasonable value of 150 psi ($\equiv 1 \times 10^6$ N m^{-2} or 10 bar) for P_2. The pressure must not of course be reduced to atmospheric by the proportioner as this will cause it to have no "throw" when released from the hose. Again, arbitrarily but reasonably we allow the pressure to drop by 10% on its passage through the proportioner in which case $P_1 = 0.9 \times 10^6$ Nm^{-2}. Using a value of 1000 kgm^{-3} for the density:

Rate of supply of energy = 100 kJ per kg of water flowed

Let the flow rate through the proportioner be m kg s^{-1}

Rate of supply of energy = 100m kW

so a flow rate of 1 kg s^{-1} gives a rate of supply of mechanical energy of 100 kW

More generally, and letting $(P_2 - P_1) = \Delta P$

Rate of supply of energy = $10^{-3}\Delta Pm$ kW

This appealingly simple expression signifies that the power obtainable is:

100 kW per bar pressure drop per kg s^{-1} flow rate.

Fluorescence, as a basis for detecting accelerants

Fluorescence occurs when a molecule accepts a photon and responds by emitting another of longer wavelength. Typically the received photon is in the ultraviolet range and the emitted one in the visible range. This forms the basis of chemical analysis and applications include the detection of accelerants where arson is suspected in the examination of the scene of a fire. Petroleum fractions such as gasoline and kerosene, acetone, benzene and vegetable oils are amongst materials used as accelerants by arsonists which will so respond to ultraviolet light. Therefore any traces of them amongst the fire debris can be detected by means of a UV lamp.

Fluoromac Plus-AR

Fluoroprotein (FP) foam. Twice the concentration for application to a hydrocarbon fire is recommended for use on a fire involving a polar liquid. It is suitable for use with dry extinguishing agents, including **Purple K**, when a "Twin Agent" approach to extinguishment is taken.

Flyash, as a component of concrete

Flyash is the solid resulting from the combustion of coal as pulverised fuel, and is frequently an ingredient of concrete. Flyash contains unburnt carbon, typically to the extent of a 10% loss on prolonged heating at 750°C. The role of the concrete in fire protection could be impaired if during an accidental fire such loss took place.

FM-200®

Chemical formula CF_3CHFCF_3, heptafluoropropane, an extinguishing agent replacing **halons** in waterless extinguishment by reason of its negligible ozone depletion potential. It is manufactured in the US by companies including Great Lakes and Dupont.

Foamousse

A family of **protein foams**, having found wide use in applications including shipping. They are capable, with suitable proportioning and generation, of moderate expansion (up to about 10, whereas some hydrocarbon-based foams such as **Jet-X** can achieve a value of 1000). The fluorine-containing member of the group is an example of an **FFFP** and termed Foamousse-Fffp. If there is additionally suitability for use on fire involving polar liquids, the name is Foamousse-Fffp/Ar.

Foams, electrical resistance of

Although foams of the **AFFF** type are not intended for electrical fires, there is always a possibility that in ignorance or in panic they will be so applied. Consequently many such foams are subjected to the "35 kV test"; to pass the test a foam has to be a strong dielectric. This is helpful not only in inadvertent use of an AFFF but also in a fire not of electrical origin in which electrical cables are present and threatened.

Foam/water sprinkler

Sprinkler systems are available which incorporate a proportioner, enabling foam to be released with the water. Such sprinklers find application to aircraft hangars and tanker loading bays; they can also be used on ships. Exampes are the K40 and K20 sprinklers from Angus Fire, which are intended for use with an AR-FFFP foam concentrate. (*See also* **WASP**.)

Fog angle

When water for fire fighting is emitted in a straight stream, the fog angle is nil. In the extreme where the water formed a barrier perpendicular to the axis of the nozzle, the fog angle would be 180°. Fog angles are important in performance and 60° is a typical value in "attack". This might be part of a **combination stream**. Much wider angles are used where fire fighter protection by means of a **water curtain** is required. A fog is required for **hydraulic ventilation**.

Fog tip

Device attached to a fire fighting nozzle to break up the water so that water is released as a fog instead of a straight stream. It is restricted to fog use only and cannot be used to convert between fog and straight stream as for example can nozzles of the **Vari-Nozzle** type. A fog tip should not be confused with a *diffuser*. The diffuser, which sometimes finds application in mild fire fighting situations, simply splits one straight stream into several by passing the water from the hose through a plate drilled with holes.

Ford, recall of vehicles by

The Ford Motor Company have since 1999 recalled over 10 million vehicles in the USA because of a fire hazard with them. The origin of the difficulty is a switch which deactivates cruise control when the driver steps on the brakes. This switch is able to access power from the battery at all times whether the engine is running or not. It is believed that brake fluid entering the circuit of which the switch is a part is the cause of the fires. The fires have tended to occur when the vehicle has been parked or garaged. The switch is estimated to have been fitted to 16 million Ford vehicles in the US.

Ford F-550

4WD vehicle adaptable to fire fighting use. An example is the Fastak, from the same manufacturer as the **Brushmaster**. Another is the Lynx, manufactured by E-One. **Sutphen** equipment has also been installed on Ford F-550 cab/chassis structures and engines fitted are sometimes from manufacturers other than Ford. (See also **East Troy WI**; **Fire appliances, used and rebuilt**; **Pinewood, NC**.)

Formtex

Manufactured by Mandals in Norway, who also manufacture **Guardman**, this hose material is intended for building use and storage as a reel when "on call". It has a woven polyester outer cover and a lining of EPDM (ethylene propylene diene monomer). The widest diameter in which it is supplied is suitable for working pressures up to about 550 psi.

Fort Collins, CO

A hotel in this locality in December 2007 was the scene of a fire in which the installed sprinkler system prevented the consequences from being more serious. Minor injuries were experienced by one hotel occupant. There was a fire at the same hotel in the 1970s before it was fitted with sprinklers: two floors of the hotel were extensively damaged.

Fort Meade, MD

Scene of delivery in late 2007 of an aerial manufactured by **Seagrave**. This aerial also has a water tank and a pump, so is presumably classifiable as a **quad** if not as a **quint**. The following month the Department also received a new pumper from Seagrave. The appliances have some structural commonality being based on the Seagrave "Marauder" frame. The Marauder can form the cab of a **tractor-drawn aerial**, as with the appliance recently delivered to **Redwood, CA**.

Fort Stikine

Vessel which, in Bombay harbour on 14 April 1944, caught fire. Its payload was very varied and had some very dangerous components: gun powder, ammunition, cotton bales and timber. There was also some gold bullion! Once the fire had begun there were two explosions. The blasts caused many deaths (>700) and injuries, and massive damage to property. There was widespread community panic in the totally unfounded belief that the incident was an act of Japanese aggression, as the Japanese were known to have entered Burma at that time.

Frangible bulb

Classical device, still in widespread use, for initiating a sprinkler. It comprises a glass vessel partly filled with liquid. Heating of the liquid breaks the glass and opens an orifice through which water can exit. An alternative is the **fusible link**.

Franklin, IN

The scene of a fire in which a two-year old child's life was saved by use of a thermal imaging camera. The rest of the family had escaped, and fire fighters went inside with their camera and were able within a few minutes to locate the child's "image". The child was barely breathing by then, and had it not been for the speedy action which the **thermal imaging** enabled, his discovery by fire fighters might well have been too late.

Frejus Tunnel

Tunnel linking France and Italy, a few years ago the scene of a fire in which two persons died. The fire began in a truck carrying tyres. One of the persons who died in the fire had abandoned his own vehicle and run about half a mile through the smoke, intending to escape from the tunnel, before collapsing.

Frothover

Mild form of boilover. Foams such as **ATC** have been effectively applied to an oil tank fire displaying frothover. An alternative is the digging of a temporary trench close to the tank or erection of a makeshift wall in order to prevent liquid having "frothed out" from spreading.

FSA-410BS

Photoelectric smoke alarm system from Norco in Riverside, CA. It has a piezoelectric **sounder** internal to the detector unit.

Fuel conductivity improver

A.k.a. a static dissipater additive, an agent added to a liquid fuel so as to increase its electrical conductivity, thereby enabling accidentally created static electricity to be dispersed and avoiding creation of a spark. Such an additive is usually mandatory for fuels for jet aircraft, and an example is **Stadis® 450**.

Fukuoka, Japan

The scene in June 1996 of an air crash involving a DC-10 operated by the Indonesian airline Garuda. A broken fuel line had led to fire during an aborted take-off. There were three deaths.

Fusible link

Access to a cabinet for flammable liquids might be required frequently and continual opening and closing might be not only inconvenient but possibly ergonomically unsound. In that event the doors can be safely latched open if a fusible link is fitted. This will melt at about 75 °C, causing door closure. The term fusible link also applies to an opening device for use with a **sprinkler head**. The cap which causes water to be retained within the head when sprinkling is not required is held in place by a piece of solder-like material, which melts when contacted by hot gas from a fire. The temperature at which this takes place can, of course, within limits, be set by adjusting the composition of the solder.

Futons, fire protection measure for

Futons originated in Japan and came into use in North America in the 1960s. Traditional Japanese futons were made entirely of cotton. Any futons on the market currently made entirely of cotton are treated with boric acid as fire retardant. Synthetic materials used as filling materials for futons include polyurethane, which can be treated with **melamine** as a fire retardant. Additionally to these chemical approaches to fire safety, a futon can contain a barrier material to raise the thermal resistance. Barrier materials include fibre glass, neoprene and cotton pads treated with boric acid. Such barrier materials were introduced into futons after it became clear that a futon once alight can have a heat release rate sufficient to cause flashover.

Fyre-Can

Cast-in-place firestop which incorporates **TREMstop WS** as the intumescent material. It comes in a range of diameters and lengths.

Fyrol™

Family of flame retardants, all phosphate based. An example is Fyrol™ 38, which is directed at the same market as **Antiblaze®195** and PUMA® 4020. Also aimed at this market is Fyrol™ FR-2 which, when incorporated into a polyurethane foam, does not lead to discolouration of the foam on heating due to carbonisation ("scorch") to the extent that Fyrol™ 38 does. There is production of Fyrol™ retardants in the US, Europe and China.

Fyroflex™

Trade name of two flame retardants for resins from the same stable as **Fyrol™**. One, denoted Fyrolflex BDP, is composed of bisphenol A bis (diphenyl phosphate) and its applications include polypropylene oxide and polycarbonates, where it has a plasticising role as well as a fire-retardant one. The other, denoted Fyrolflex RDP, is composed of resorcinol bis (diphenyl phosphate). It finds applications to polymer blends. A liquid at ordinary temperatures, it is incorporated into a polymer melt as such enabling the rheology as well as the propensity to fire to be controlled. **Triphenyl phosphate** is also a fire retardant.

G

Garibaldo

New tug vessel built in China recently sold to a European owner having changed hands in Hong Kong and taken from there to the Tyne in north east England. She is equipped for fire fighting at **FiFi I** level, having two **monitors** each capable of 1550 m³ per hour delivery and 100 m "throw". The vessel carries 18 m³ of foam concentrate.

Gas chromatography, use of in arson detection

Accelerants such as kerosene in the debris from a fire are evidence of arson. Amongst the methods for detecting such accelerants is gas chromatography. A remnant of the fire remains, such as a piece of carpet, is heated in a jar to about 100°C, and a sample of the gas in the headspace syringed into a gas chromatograph with (usually) a packed column and a flame ionisation detector. A substance such as kerosene will give a "fingerprint" pattern on the resulting chromatogram across a short range of residence times.

Gasoline, use of by arsonists

Partly because of its very easy availability, gasoline is a common choice of accelerant in arson attacks. In fact it is the accelerant most frequently identified by forensic laboratories in the US. The next material up in the boiling range of crude oil is naphtha and this too has been used by arsonists. Most of the naphtha at a refinery is further chemically processed by reforming or cracking but some unprocessed naphtha is sold for solvent purposes, also for camp stove fuel.

Gassonic

Ultrasound gas detector, finding particular application to hydrocarbon gases. Its resolution is such that it can detect a leakage of 0.1 kg s⁻¹ at a distance of 12 m.

General Kelley

Name of the fire fighting vessel which serves the Port of New Orleans, having played a role in attempts to extinguish the fire on the **Balmoral Sea**. About five years later it was involved in the waterfront fires that followed Hurricane Katrina, in one of which there were exploding vessels of propane. The vessel has a crew of nine.

General Slocum

Steam ship, launched in 1891 which caught fire whilst in service off New York in June 1904. The death toll was over 1000. On the day of the accident she had been chartered by a local church to take its members on an outing. The vessel was carrying flammable liquids. The ship's fire rescue facilities, including the hoses, were in poor condition and not working properly, nor had the crew of the vessel ever rehearsed a fire drill. It is believed that cork blocks for use in such an emergency ("life preservers") had been partly filled with metal to raise the ship's weight and this, of course, totally negated their potential value. There were many indictments (in the formal sense) resulting from the accident and the Captain of the vessel (who had lost the sight of one eye as a result of the fire) received a prison sentence. One of the survivors of the accident was a six-month old baby Adella Liebenow who, as Adella Wotherspoon, died as recently as 2004 in her 101st year.

Genesis

Passenger vessel renamed *Oasis of the Seas* which was built in Finland and which, upon delivery was the largest such vessel in the world with the capacity to carry 5400 passengers. It was to be fitted with a **Hi-Fog®** system for fire protection, there being in excess of 13 000 outlets (nozzles) across the entire vessel. There will be a greater concentration of these in such parts of the vessel as the engine room and laundry. Installation of the Hi-Fog® plant was concurrent with the building of the respective parts of the vessel, in preference to its fitting all at once on completion of the ship. This has two advantages. First, it makes for integral design of the fire protection system. Secondly, and perhaps more importantly, *the vessel as it takes shape will be fire protected during building*. This will prevent a repeat of the events preceding the launch of the **Diamond Princess**.

Geonisio Barroso

Anchor handling, tug and supply (AHTS) vessel in Brazilian waters, entering service in May 2004 and operated by Delba Maritima. For its role in fire fighting it has two **monitors,** each with a delivery capability of 3600 m³ per hour qualifying the vessel for **FiFi II** classification. The *Geonisio Barroso* was one of three vessels from the same shipyard supplied to Delba Maritima between May and October 2004. She is better equipped than either of the other two (whose names are *Haroldo Ramos* and *Yvan Baretto*) in fire fighting terms having, in the combined and complementary operation of the trio of vessels, been assigned the fire fighting role.

Gerda Maira

German trawler upon which, in August 1999, there was a fire. The trawler was in UK waters off the Norfolk coast at the time of the fire. RAF helicopters were used to winch the crew of the trawler to safety. Aid was also received from a North Sea ferry and from a French shipping vessel which towed the *Gerda Maira* out of UK waters into Dutch waters where the Dutch authorities took over.

Glasgow Airport, terrorist attack at in 2007

On 1 July 2007 a 4WD vehicle containing LPG cylinders rammed into the terminal building at Glasgow Airport and caught fire. The author had significant media involvement in the matter and made the point that possible danger was from fireball behaviour had the LPG cylinders leaked and their contents ignited and from flashover had the vehicle entered the terminal hall.

Glenpool, OK

The scene of a fire at a tank pool in 2006, which began as a result of a lightning strike. The affected tank was holding gasoline and had an **internal floating roof**. There was no spread to nearby tanks. Removal of fuel from the tanks via pipes from it took place during the fire and diverted either to other tanks at the site or to distant locations by pipeline. As well as mitigating the possible effects of the fire, this had the benefit that most of the fuel in the tank was retrieved and financial losses thereby reduced. There had been a fire in 2003 at the same site in Glenpool at a tank owned by a different operator. Static electricity is believed to have caused this.

Glotek™

Jacketed fire hose having an outer protective layer ("encapsulation") containing **Encap™**. Like its stablemate **Tidalwave 600™** (both are Tyco products) it has a thermoplastic polyurethane (TPU) lining material. It is capable of operating at up to 800 psi.

Gloucester, MA

The scene in December 2007 of a fire in an apartment building having the high value of 8 on the **alarm number** scale. There was one fatality and destruction of a Jewish temple nearby.

Godiva World Series

Range of **vehicle-mounted pumps** for fire fighting, of centrifugal design as almost all such pumps now are. To its use in city and municipal service have been added a number of specialised applications, of which two will be noted. One is its use in remote parts of the Czech Republic where the terrain is inhospitable making access of emergency vehicles difficult. Another is its use in fire apparatus expressly designed and built for use in Alpine tunnels.

Goettingen

This German city was recently the scene of a fire in which a fire fighter was killed. When his body was found it was evident that he had removed his **SCBA** and an examination of it indicated a difficulty with the air supply. There was a detailed response from the manufacturer of the SCBA.

Golden Age Nursing Home fire

This building in Fitchville, OH was the scene of a fire in November 1963 in which 63 elderly people died. Initial attempts to contact the fire department failed because the telephone wires had been burnt during the early stages of the fire. A truck driver passing by brought the fire department, and he and two other truck drivers helped the fire fighters rescue the 24 survivors. Their task was made more difficult by powerful winds. It is also believed that when the alarm was raised the residents had made for their rooms instead of going outside to safety. The building had in fact passed a fire safety inspection eight months earlier. It was the worst fire in the US in terms of loss of life for almost five years. Coverage of it was, however, diminished by the fact that it occurred less than 24 hours after the assassination of President John F. Kennedy.[18]

Goodge Street

Station on the London Underground where in June 1980 there was a fatal fire. The ignition source was a discarded cigarette and those campaigning

---

18   The **Hotel Roosevelt fire**, which is also seen as one of the worst fires in the US during the mid-20th century, was only about a month later.

for a ban on smoking on the Underground had a stronger case as a result of the tragedy. Such a ban actually came into effect four years later.

Good Will Fire Company

In US fire fighting terminology "company" means "group" or "team", usage which does not occur in English-speaking countries such as the UK and Australia. Some volunteer fire fighting teams in the US call themselves Good Will companies, identifying themselves by following that term with the name of the place at which the company is based (plus a numeral preceded by a # if there is more than one Good Will company in one particular town or city). Being volunteer companies, these are less well funded than companies associated with city fire departments. They therefore tend to go for second-hand appliances often extending their useful lives by many years by suitable refitting (as at **Jacobus, PA**). One exception is **Darby Township, PA** where a new appliance was delivered in 1998. Similarly, it was a major achievement in fundraising terms when recently the Good Will Hose Company of Cumbola, PA took delivery of a *new* 1500 gpm pumper manufactured by **Sutphen**. In black livery, this pumper is equipped with 1600 feet of hose of various specifications. The Good Will Hose Company of Cumbola protects an area of about 10 square miles within rural Pennsylvania. A Good Will Company might have "career fire fighters" as well as volunteer ones (as at **Newcastle, DE**) in which case such fire fighters are employees of an organisation supported at least in part by voluntary subscription. A particular fire might be attended both by a Good Will Company and a city fire department where their combined effort is required ("mutual aid").

Gothenburg disco fire

The death toll was 63, and there were over 200 injuries, as a result of this fire in Sweden in 1998. All of the dead and injured were in the age range 12 to 20. The fire was started by a group of teenagers who had been required to leave the premises for misbehaviour. They faced criminal charges and received jail terms.

"Gravity-Nurse"

Modus operandi whereby extinguishing water from a **water tender** is passed along **hard suction hose** to the **attack apparatus** under gravity and without an intervening **dump tank**. A tanker operating in the gravity-nurse mode will itself be replenished with water.

Grayson, KY

One of a good number of "communities" in the US having a split rating on the **Public Protection Classification (PPC™)** scale, denoted 6/9. A property in Grayson within five miles of a fire station and 1000 feet of a hydrant is class 6. All other properties in the area to which the Grayson Fire Department responds are class 9.

Great Chicago Fire

Occurring on 8 October 1871, the death toll being 300. Cause unknown, although in folklore it is said to have started in a farm shed. Chicago at that time had many wooden buildings, and a more than usually dry summer in 1871 had caused the moisture content of the wood to drop making such buildings more susceptible to fire. On the very same day (a Sunday) major fires also occurred at **Peshtigo, WI** and at Port Huron, MI.

Greenwich Village, apartment block fire 2004

This incident was initially classified 3 on the **alarm number** scale and later, as more help was sent for, reclassified as 4, denoting the participation of 168 fire fighters with 39 vehicles. The apartments were tenant occupied and according to one account complaints about their condition had been made to the owner. Three fire fighters were hospitalised.

Ground ladder

Lowest level of ladder in direct contact with the ground unlike a roof ladder or an aerial ladder. A ground ladder may be a **fibre glass ladder** or it may be constructed wholly of aluminum or wholly of wood.[19] Advantages of wood include its thermal insulation properties. Rungs on such a ladder will be solid wood, often more robust than the hollow rungs on an aluminium ladder. On the negative side, timber is less intrinsically uniform in its mechanical properties than aluminium, making careful selection of raw materials necessary. An aluminium rung if broken can be replaced *in situ* during fire fighting, whereas a wooden one cannot. An aluminium ladder offers no protection at all from **fire ground electrocution**.

GRP

Glass fibre reinforced plastic. Such materials find application in both active and passive fire protection. An example of the former is their use in **water tenders**: an example of the latter is their use in the fabrication of doors.

---

19 Some wooden ground ladders in use with the San Francisco Fire Department have been in service since 1919 having, of course, had repairs and maintenance over that time.

Guardman

Fire hose composed of **nitrile** with a woven textile out layer, available in diameters up to 6 inches and capable of high operating pressures. It is used by fire brigades and also finds considerable marine application, being approved by Det Norsk Veritas (DNV). It is manufactured by the Norwegian concern Mandals.

Gulf Coast states, fire death rates of

In units of deaths per million population in 2004, these are Texas 10.7, Louisiana 21.3, Alabama 25.6 and Mississippi 32.1. With the exception of Texas these are well above the national value of 12.4 and Mississippi has the highest value of all of the states (although that for the District of Columbia is higher still). The **Public Protection Classification (PPC™)** profile of Texas is very strong, there being several class 1 communities.

Gyproc

Board material for use in buildings, containing **gypsum** and manufactured by British Gypsum. From the same manufacturer comes Arteco, a range of tiles for ceiling use also based on gypsum which have found use in buildings including hospitals and schools.

Gypsum

Mineral of chemical formula $CaSO_4.2H_2O$ finding wide application as the major component of materials for fire-resistant structures. Evaporation of the water of crystallisation retards uptake of heat by gypsum and this is an important factor in its effectiveness. Gypsum alone lacks the mechanical properties usually required in building and has to be reinforced. Glass fibre is sometimes used for this purpose. Gypsum is also used in fireproofing substances applied by spray or by injection during building. It is a component of **Monokote®**, the most widely used such preparation in the world.

Hi-Fog® water mist system
in action (courtesy Marioff
Corporation Oy)

H

Habo Church

Located near Jönköping in Sweden and built in 1723, this building has many very valuable paintings and carvings on its inside walls. Vulnerability to fire is high as the church is made of wood. There is now a **Hi-Fog®** protective system in place. The minimising of water damage in the event of fire, an advantage of Hi-Fog® over conventional sprinklers, is of particular value in this case where the art treasures are integral to the walls and not removable.

Hadi IX

Offshore supply and maintenance vessel with FiFi ½ (≡ **FiFi 0**) facilities, having one pump and two monitors, total output 1200 m³ of water per hour. The vessel was built in Hong Kong and is owned by Hadi H Al Hammam, who have an extensive fleet of vessels servicing offshore facilities in the Middle East including the *Hadi XI*, an AHTS vessel built in Singapore. Like very many AHTS vessels, Hadi XI is **FiFi I**.

Hair spray cans, fire and explosion incidents with

An example of such an incident occurred in Hampshire, UK in 2006, when inside a car which had become warm through sunlight a can of hair spray exploded and injured the driver of the car. The other two occupants of the car (one of them a baby) were unhurt. A photograph of the accident scene shows destruction of part of the upholstery of the car. A hair spray is likely to contain a hydrocarbon such as propane or butane as propellant and ethanol as solvent.

Hakodate City

In the Hokkaido Prefecture of Japan (sometimes called "Japan's New England"), the scene in 1934 of a huge fire. At the time that the fire began there was a 70 mph gale and this obviously aided the spread of the fire. There were 1500 dead and 23 000 homeless survivors, some of whom sought temporary accommodation with local Ainu people.[20]

---

20 The so-called aboriginals of Japan.

Hale Typhoon®

Range of fans for **PPV**, from Hale Products. These are powered not by an engine but by a water turbine. The water turbine has in fact found application to other facets of fire engineering including the driving of a stirrer for foam preparation.

Halogen lamp

This comprises an incandescent light using a tungsten filament, surrounded by a quartz tube containing a halogen. It can operate at higher temperatures than in the absence of the halogen as it provides for return of any tungsten lost from the surface. This is brought about by formation of tungsten halides, which on contacting the tungsten surface decompose. Halogen lamps find application to emergency lighting.

Halon 1211

This is bromochlorodifluoromethane, CF_2ClBr. The boiling point of this substance is $-4°C$, and in fire extinguishers it is contained at room temperature under its own vapour pressure, and emerges from an extinguisher as a liquid under non-equilibrium conditions. The vapour pressure inside the extinguisher is fairly low and it is usually necessary to raise the internal pressure by inclusion of nitrogen in order to achieve a sufficiently rapid supply in application to a fire. Halon 1211 has been widely used to very good effect in aircraft.

Halon 1301

This is bromotrifluoromethane, $CBrF_3$. Its boiling point is $-58°C$. It is contained in extinguishers as a liquefied gas having an equilibrium vapour pressure at $25°C$ of 16 bar. Unlike **Halon 1211** it exits an extinguisher as a gas, and in choice of one or the other the fact that one is a liquid stream and the other a gas is often a decisive factor. The two have in fact often been used as a blend. An obvious advantage is that the vapour pressure of Halon 1301 eliminates the need for nitrogen with Halon 1211 and so replaces a merely inert propelling gas with one having active extinguishment potential.

Halon 2402

This is dibromotetrafluoroethane, CBr_2FCF_3, boiling point $47°C$, therefore applied as a liquid. Its superior performance to **Halon 1211** and **Halon 1311** in some applications has been attributed to the fact that it contains two bromine atoms in each molecule. Before the prohibition of **halons** there was a patent application for a device whereby Halon 2402 would be contained in a thin-walled glass container as a fire protection device disguised as a Christmas tree decoration. In the event that the tree caught fire, the vapour pressure of the Halon 2402 would build up sufficiently to break the glass and release the contents.

Halons

Generic term for a group of fairly simple organic compounds used as fire extinguishing agents. Particular examples have entries of their own above. Halons have found wide application as extinguishants throughout the second half of the 20th century, for example in aircraft hangars and on ships. They have also been used at offshore oil and gas facilities. There has been a strong move away from halons in more recent years because of their effects on the ozone layer. International coordination of ozone layer protection began in 1987 with the Montreal Protocol and in the US all manufacture and import of halons ceased. Existing stocks remain in use and recycled halons are available for the recharging of extinguishers.

The EU has exceeded its obligations under the Montreal Protocol by proscribing the use of halons except in certain specified circumstances and requiring decommissioning of extinguishing systems using halons. Halons are still produced in countries including South Korea, China and the former Soviet Union.

Halotron™

The prohibition of **halons** as extinguishing agents does not preclude all totally halogenated hydrocarbons from such use. The extent to which a particular one will deplete ozone after use can be determined by a standard test for ozone depletion potential (ODP). Such substances having an extinguishing capability comparable to that of halons but a lower ODP are therefore coming into use as halon substitutes. Halotron™ is one such. It consists of 90% 1,1,1-trifluoro-2,2-dichloroethane, which in its chemical structure differs from **Halon 1211** and **Halon 1301** in two important ways: it is C_2 not C_1 and is not quite entirely halogenated, there being one hydrogen atom in the molecule. Whereas in the arbitrary but soundly based units of ODP Halon 1211 has a value greater than 4, that of Halotron™ is 0.014. **FM-200®**, another halon substitute, has an ODP off-scale on the low side and therefore for practical purposes zero. Another halon substitute is **iodotrifluoromethane**.

Hamlet, NC

The scene in 1991 of a fire at a chicken processing plant in which 25 workers died and 49 more were injured. The fire began when hydraulic oil leaked and was ignited by a fryer. Employees were unable to escape because the exits had been locked. Criminal charges resulted and the owner of the plant received a long custodial sentence. It was noted that during the 11 years of the plant's existence it had not had a safety inspection. The fire has, not unreasonably, been compared to the **Triangle shirtwaist factory fire**, from which it is separated in time by 80 years.

Happy Land Social Club

In New York's Bronx area, the scene of a fire in 1990 in which 87 lives were lost. Over a year earlier the club building has been declared unsafe, so the 1990 event at which the fire occurred was irregular. A Cuban refugee set fire to the premises using gasoline as an accelerant. The fire exits from the club had been locked to stop people from using them to avoid the entry fee. The sentence the perpetrator received on multiple counts of murder and of arson was the longest ever imposed in the State of New York. The street in which the club stood is now called the Plaza of the 87, as a memorial to the victims.

Hard suction hose

This has an internal helix of steel. The capability of a hose to flatten when out of service does not of course extend to these which, in addition to the steel, might have a layer of PVC on top of the woven layer adding still more to the rigidity.

Harlow, Essex

Scene in 2007 of a fire at an in-house bakery at a Tesco store necessitating evacuation of the store. An electrical fault is believed to have been the cause.

Hartindo AF11E

A halon substitute for extinguishment purposes having the same primary constituent – dichlorotrifluoroethane – as **Halotron™**. It is manufactured by Newstar Chemicals in Malaysia and the proprietary gas which complements the extinguishing agent is of their development. Their other products include **Hartindo AF21** and **Hartindo AF 31**.

Hartindo AF21

Water-based flame inhibitor for application to furnishings, carpet and clothing. When applied to such it leaves a residue which will char under the influence of heat and so restrict combustion. Marketing of this and other fire protection products in the Hartindo family of products in the US is by the MSE Enviro-Tech Corporation.

Hartindo AF31

Water-based fire extinguishment agent introduced in April 2007. It is intended for a variety of applications but believed to be especially suitable for forest fires, to which it can be directed from the air.

Hartsfield International Airport

This airport in Atlanta, GA was the scene in June 1995 of a fire involving a DC-9 aircraft. The airport had recently acquired a **Colet Jaguar** K/15 **ARFF** vehicle, at that time one of very few Jaguars in service. This reached the DC-9 *less than one minute* after the alarm was raised, and foam was directed at the fire which was extinguished within 15 minutes of the alarm.

Hasik

Tug vessel manufactured in the Netherlands having recently entered service. It has two water pumps capable of releasing in fire fighting 12000 m³ per hour of water, with or without foam. The vessel consequently has **FiFi I** classification. It has supplementary water sprays for protection of the vessel itself when it is being used in fire fighting mode.

Haslingden, Lancashire

The scene in November 2000 of a major blaze at a factory where carpet underlay is made. The fire was attended by 150 fire fighters and, interestingly, employees of the factory were asked by the fire services to assist them. Such help included the removal by forklift truck of stacks of underlay in order to create a fire break. The practice of using supervised employees to help fire fighters was novel in the UK at that time although it had been followed in the US.

Hassan Taha

Vessel operating on the Nile whose sole purpose is fire fighting. Its *raison d'être* is the need for such a facility not only for cruisers carrying tourists but also for the numerous "floating hotels" on the Nile. There are two pumps, and the total supply rate of water is 960 m³ per hour. This is below the level required even for **FiFi I** classification but such a perspective fails to compare like with like as the *Hassan Taha* is not concerned with offshore operations involving hydrocarbons. The vessel carries foam.

Hawaii, participation of in the Federal Excess Property Program

There has been extensive use of vehicles acquired through the Program in Hawaii, and a recent report notes that such vehicles are by now very elderly and difficult to maintain. The State has a plan to acquire at least one new fire appliance of its own each year over the next decade. The other of the states not in the "lower 48" – Alaska – has also benefited in its fire fighting capability from the **Federal Excess Property Program**.

Hawaii Mars

Flying boat in service in British Columbia as an **air tanker** ("water bomber") with the sister flying boat *Philippine Mars*. Each was initially manufactured in the 1940s and purchased for fire fighting use in BC in 1959. Each is still in use and based on Vancouver Island. Each can carry about 7000 gallons of water, about twice the capacity of a **P3 Orion** or a **C-130 Hercules**. The *Hawaii* and the *Philippine* are however differently configured, the former having water tanks where fuel tanks originally were and the latter having them where the cargo hold previously was. The substitution of fuel tanks for water tanks with the *Hawaii* means that it can hold only about half as much fuel as the *Philippine*. Each carries foam concentrate. When the water tanks need refilling, it is obviously from the stretch of water on to which the sea plane has descended and this makes for very rapid filling. Adding to the sense of time warp in fire fighting in BC is the fact that many **Convair 580** aircraft are in use there.

Hawiyah

Scene of a natural gas liquids (NGL) plant in Saudi Arabia where in November 2007 there was an explosion killing 28 people, the majority of them expatriate workers.

HC-12a®

Refrigerant, having found a role as a replacement for Freon, CF_2Cl_2. HC-12a® is a blend, not a pure chemical substance, and contains significant amounts of propane and butane. DURACOOL 12a® is an equivalent product from a different manufacturer.

Heart and Soul Restaurant

The creation of this eatery in Liverpool, England out of a derelict building first erected in the 1770s required fire protection measures, with regard to the hazards of ovens, hotplates and other kitchen plant. Accordingly **Pyroguard** glass of 7.2 mm thickness was installed in parts of the restaurant including the entrance and the partitioning between the different parts of the restaurant.

Heavenly bamboo

Botanical name *Nandina domestica*, this species is indigenous to Asia and is often encountered in gardens in Japan. It has been naturalised in other parts of the world, including North America, and in California is used as a **firewise plant**.

Helitorch

Device carried by a helicopter whereby a substance similar to **napalm** for **controlled burning** of forests and grasslands is carried and released when it is required. The device (which might sometimes be carried by a small fixed-wing aircraft) is a drip burner; napalm exiting it is ignited before its descent. One of the most recent endeavours with the helitorch to have become world news was in western Michigan, where a helicopter crew on "loan" from Ontario attended a forest fire. The helitorch was used to remove fuel in the path of a rapidly propagating fire, thereby limiting its extent.

Henri Bourassa

Station on the Montreal metro system where, in December 1971, a train collision was followed by an electrical fire. There was one fatality. This and another electrical fire on the Montreal metro two years later, in which there were no fatalities, caused the loss of 45 railcars. Four more were lost on the metro system of another Canadian city – Toronto – in 1976. This was the work of an arsonist.

Hester Hall

Dormitory at Murray State University in Kentucky, where in September 1998 there was a fire which killed one student resident and injured several others. Nine months later a student resident of the dormitory was arrested on a charge of arson and murder. He had himself been removed from the building by fire fighters using an aerial ladder. The dormitory was not served with sprinklers.

Hexabromocyclododecane (HBCD)

An example of a brominated flame retardant, having molecular formula $C_{12}H_{18}Br_6$. It finds major applications in polystyrene foam and extrusion, where incorporation in a quantity of less than 1% provides an acceptable level of protection. It is also applied to textile fibres, for example in soft furnishings. There is manufacture in countries including the US, the Netherlands, Israel and China. HBCD is also made in Japan by the Jiang Yin Trustwell Chemical Co., which produces about a fifth of that country's HBCD needs, the remainder being imported. Given the scale of HBCD usage, its health and environmental effects have had to be assessed. It is believed not to put users at risk in any way and not to endanger workers handling it provided that routine protective measures are taken. Only traces of HBCD have ever been found in soil, though such trace amounts appear to be on the increase in the UK.

HFC-32

A.k.a. R-32, refrigerant substance, difluoromethane CH_2F_2. Not being totally substituted with halogen atoms it burns, according to:

$$CH_2F_2 + O_2 \rightarrow CO_2 + 2HF$$

Interestingly, its flammability range in air is wider than that of methane itself, which is 5–15%. The flammability range of HFC-32 in air is in the approximate range 14–30%, "approximate" because independent experimental determinations have imposed considerable error bars on both upper and lower values.

Hi-Fog®

A product of Marioff in Finland, Hi-Fog® combines the features of a sprinkler system with those of a **water mist fire extinguisher**; that is it provides water droplets in the sub-millimetre size range but from pipe work supplied with water by a pump, not from a singly charged cylinder. A Hi-Fog® device is initiated like a conventional sprinkler, but water on exit is dispersed into small droplets instead of simply being deflected as with a regular **sprinkler head**. Hi-Fog® was developed initially for protecting ships but has since found numerous other applications, including the **Cirrus Building**, the **Duchess Anna Amalia Library, Dulles Airport, Habo Church, La Scala**, and the **M30 Tunnel**. The London Underground and the Madrid Metro both use Hi-Fog®. (*See also* **Genesis**; **Wallenius Lines**.)

High expansion foam

Such a foam is of special value where a fire is in an enclosed space so that continuous application of an extinguishing agent might lead to flooding. A basement fire is an obvious example. High expansion foam concentrates are used at levels of between 1% and 3% in water and pass into a high expansion foam generator driven by a water turbine, requiring no electrical power. (This works similarly to hydroelectric generation of electricity, kinetic and potential energy of the water being the origin of work at the turbine blades.) With some high expansion foam concentrates, including **Jet-X**, expansion ratios of up to 1000 are routinely available by this means.

"Highly flammable tree"

The term is of course relative; there is no such thing as a non-flammable tree. However, some have been observed to burn more rapidly than others and are therefore deprecated for garden use. The term "highly flammable tree" has been coined and an example is the California laurel, botanical name *Umbellularia californica*. Other examples include **bald cypress, Monterey pine, knobcone pine** and **tan oak**.

High-volume hose

It has been described in other entries how nozzles and **monitors** have to conform to volumetric release specifications. By continuity, mass release rate at a nozzle is equal to the mass flow rate through the hose which is supplying it. Since water is an incompressible fluid, the volumetric flow rates are also the same. High-volume hose is available with inner diameters in the approximate range 4-6 inches. These have structures and configurations like those of narrower hose, comprising a synthetic rubber interior and a woven fabric outer layer. Such a hose of 6 inch diameter and 100 m length would have a weight in the neighbourhood of a fifth of a tonne and would be able to withstand about 40 bar internal pressure. A difficulty with such hose is that it will not, when out of service, lay flat, so storage spaces required are significant. There are variants which have smaller wall thicknesses of the layered materials whilst retaining the same inner diameter and so *will* lay flat: the down side is that these are limited to lower pressures than the more robust type.

Hino

Truck manufacturing company, a division of Toyota. There has been collaboration with the Morita Corporation in the building of fire appliances. These include a turntable ladder appliance (TL) recently delivered to the Tokyo Fire Department. There are Hino fire trucks in service in other places, including the Australian state of Victoria, where a 650 gpm Hino pumper has been in use with the Country Fire Authority since 1985.

HMCS *Chicoutimi*

This submarine, which was not nuclear powered, was originally built for the Royal Navy. After a period of mothballing she had a refit, and was bought by the Canadian Navy in October 2004. Shortly after she had departed from a base in Scotland on the delivery voyage to Canada there was a fire on board as a result of which one crew member lost his life through smoke inhalation. An analysis of events is as follows. An air vent was being repaired, and whilst engineers were working on it authority was given for the hatches to be opened. There was a surge of about 2 tonnes of seawater into the vessel and an electrical fire followed from it. The captain was, in the follow-up, vindicated in his decision to allow the hatches to be opened whilst the vent was being repaired. Had a risk analysis been done on the vent repair operation, the frequency of ingress of a quantity of water of that magnitude would have been exceedingly small and the term "rogue wave" used in the follow-up report expressed that view less formally. HMCS *Chicoutimi* was sea-lifted to Halifax, Nova Scotia, having been welded to a Norwegian vessel, and will return to service in *circa* 2010. There was a little acrimony between the two countries as a result, partly because of a statement by the UK Secretary of State for Defence that Canada ought to have taken a "buyer beware" attitude.

Homes for the elderly, legal responsibilities of operators of

Two recent prosecutions of care home operators in the UK will be invoked as illustrations. The owners of Ravenscroft Park Nursing Home in Barnet, London were fined £200 000 as a result of a fire at the home. Issues raised in the prosecution included the fact that a bedroom door was locked whilst the occupant was inside. Fire exits had been secured with padlocks, and a corridor upon which escape from fire depended was being used for storage. The matter of the adequacy of the risk assessment for the home was raised. Prosecution also followed when at a home for the elderly in Rhos on Sea, Wales, there was a fire which began in the laundry room. Fire doors had been wedged open, necessitating rescue by the fire services of four residents who would otherwise have been protected by the fire doors.

Hong Kong, two fatal fires in 1996

There were two fatal fires in Hong Kong in 1996, the year before handover back to China. The first, in February that year, occurred in Pat Sin Leng Country Park in the New Territories. A school group numbering 54 were hiking in the Park when they encountered a rapidly propagating grass fire. The death toll was four – two teachers and two students – and there were 14 serious injuries. In November of the same year a fire in a 16-storey commercial building in Hong Kong claimed 40 lives. The building was undergoing some refurbishment, including the installation of new lifts, and there was bamboo scaffolding in place. All of the fatalities except one occurred in the upper storeys. Welding was taking place at one of the upper storeys and this is believed to have provided an ignition source.

Hopcalite

So-called carbon monoxide filter for use in breathing apparatus worn by fire fighters and also commonly incorporated into **escape masks**. It does not in fact filter, but catalyses the conversion of carbon monoxide to carbon dioxide. Its principal constituents are oxides of manganese and copper.

Hose, maintenance of

Fire fighting hose is a significant capital asset of any fire department and is in no sense an "expendable". It is therefore important that hose returning from duty is cleaned before being put into storage to await further duty. This is brought about by vigorous cleaning of the outer layer with water, and several devices are available for this. One such, having been connected to a water supply at about 40 psi, receives the hose via an automatic feeder, so that the previously dirty hose emerges clean and, after drying, ready for return to storage and to standby status. Such drying might well be natural, although low-temperature natural convection heat enclosures are in fact available for this purpose. If it is desired to remove air and water from the inside of a hose, it can be passed through rollers. Hose composed of **textile-reinforced rubber**, of which **Storex Laylite** is an example, does not require drying after washing and this is a significant benefit of such hose.

Hose stream test

Once a glass product has been tested for its ability to withstand heat from a fire for 20 minutes or more, it can be subjected to the hose stream test. This simply means that whilst still hot it is sprayed with water from a hose, the pressure of water within which is at least 2 bar. In the USA only a glass which remains intact during the hose stream test can be declared to have a "fire rating" of 20 minutes or more. In Canada the hose stream test has to be applied even for a fire rating of less than 20 minutes.

Hotel du Vin

The opening of this luxury hotel in Newcastle, England was delayed as a result of a fire in the partly complete building in November 2007. A few weeks later there was a fire at the Terra Vina, a hotel in the south of England having a degree of co-ownership with the Hotel du Vin. This started in the hotel's kitchen and was clearly accidental. The damage was much less extensive than that at the Newcastle hotel, which might have been the work of arsonists.

Hotel Roosevelt fire

The Hotel Roosevelt in Jacksonville, FL was destroyed by fire in December 1963. A new ceiling had been installed but the exisiting one, which had been seen as a fire hazard, was not removed. This is believed to have provided fuel for the fire once it had been started by an electrical fault. Twenty-two residents of the hotel were found dead in their rooms from carbon monoxide poisoning. One of the fire fighters died of a heart attack quite soon after arrival at the scene.

Hotel Winecoff

Situated in Atlanta, GA, this was the scene in 1946 of a fire in which 119 persons were killed, including a few fire fighters. There were 280 persons in the building at the time the fire started. The design of the building, which was such that it had a square cross section was one reason for the tragic consequences. The elevator system was at the very centre of the square, creating two hazards: the action of the elevator shaft in ventilating the fire and its distance from exit points. The height of the building exceeded that which the rescue ladders carried on the fire trucks could reach and the trucks themselves were prevented from getting close to the fire by the narrowness of the thoroughfare at one side of the hotel.

Houses in Multiple Occupation, susceptibility to arson attacks

Houses in Multiple Occupation (HMOs) means houses occupied by more than one independent household and is a recognised term in law relating to the renting of residential property. In arson statistics, HMOs stand out from other types of dwelling by their vulnerability. Arson attacks at HMOs are often a reaction to disagreement amongst occupants asserting (sometimes in different languages) their respective rights. Communal housekeeping is not expected to be of a high standard in HMOs, and a pile of rubbish on a landing or in a stairwell might be a temptation to an aggrieved tenant in a destructive frame of mind.

Huey

A.k.a. the UH-1 Iroquois, helicopter range manufactured by Bell in Fort Worth having found wide application to aerial fire fighting. There have been extensions to and variations on the basic design over a long period of manufacture. A large number of Huey helicopters were made for the Vietnam War, 2500 of which were lost. Some of those which made it back were sold and subsequently used for fire fighting and more recently many have been built expressly for fire fighting. The California Department of Forestry has nine "Super Hueys" in service at the present time each of which can carry 360 gallons of water and can fly for two hours without refuelling. The Los Angeles Fire Department also operates them. Other parts of the US in which they are used in fire fighting include Montana, Oregon and Washington.

A number of Hueys each fitted with a **Bambi bucket®** have recently been purchased by South Africa for use in veldt fire control.

Hurst, TX

Scene in September 2005 of a fire in a clothing shop in a shopping mall. The local fire department had no difficulty extinguishing the fire but requested the help of the Dallas/Fort Worth Airport fire services in controlling the smoke, which by entering other parts of the mall threatened lives and property and required **positive pressure ventilation** for removal. A mobile ventilation unit (MVU) was sent from the airport to the mall. The ventilator was positioned at the entrance to the mall and after half an hour of operation had caused the pressure within the mall to exceed that of the surroundings sufficiently for smoke to exit the mall in response to the pressure differential. Even during the initial pressurisation process there was significant improvement in visibility, presumably because of smoke dilution. Most of the corridors and shop interiors were clear of smoke an hour later and four and a half hours after engagement of the MVU ventilation of the mall was total. All of the shops except the one where the fire had started were open for business the following morning!

Hydraulic ladder

Alternative to a TL and equivalent in function, hydraulically operated structure having several sections moveable independently from a mounted control in order that a target rescue point be accessed or **monitors** be operated. In applying such terminology it should not be forgotten that a TL (e.g. that manufactured by **Volkan**) often works hydraulically. For this reason a hydraulic ladder might be better referred to by the approximate synonym hydraulic platform. An appliance having a hydraulic ladder instead of a TL and qualifying in other respects can be referred to as a **quint**. The hydraulic ladder is in some quarters becoming favoured over the TL in the early 21st century, for example the Merseyside Fire Service in England has gone over entirely to the hydraulic ladder and currently operates no TLs. A hydraulic ladder is sometimes referred to as a "snorkel".

Hydraulic oils, fire hazards with

Performance criteria for hydraulic oils relate largely to viscosity and, very importantly, the variation of viscosity with temperature. Fire hazards with such fluids are assessed by means of their flash points. Hydraulic oils are made from diverse sources, often from crude oil, as with **Mobil Aero HF**. There are also vegetable-based hydraulic fluids such as **BioSOY™**, **EnviroLift™**, **Hydro Safe®** and **NATURoil**. The flammability of hydraulic oil was a major factor in the aircraft accident at **Blossburg, PA** and in the 2007 freeway incident in **Mission Hills**. (See also **Hamlet, NC**; **Liège, fire in home for the elderly**.)

Hydraulic ventilation

Smoke removal from inside a building, taking place when extinguishment is complete or close to complete. Smoke is removed by application of water in the form of a fog close to an opening. Smoke is drawn into the flow so generated and exits the building.

Hydromulching

After a forest fire the ground itself over which the fire propagated needs repair. This can be brought about by application from the air of hydromulch, which resembles in composition conventional wood mulch but also has a chemical agent ("tackifier") to enable the mulch and the soil to which it is applied to become consolidated. The **Fire Boss** is a suitable aircraft for hydromulching operations.

Hydro Safe®

Hydraulic fluid based not on crude oil but on rapeseed. From the fire safety point of view it is comparable to most conventional hydraulic oils, having flash points, depending on the precise specification, in the 220 to 260°C range.

Hypalon®

Elastomer composed of chlorosulphonated polyethylene. A DuPont product, it has found application to fire hose, including the **DarQuest Lightweight** fire hose, as the outer coating material, where it shows good abrasion resistance and imperviousness to water. Amongst some hose manufacturers it has been replaced by **Encap®**.

Hythe Marina

At Southampton, England, the scene of a fire in 2005 where two small boats were destroyed and four persons had to be hospitalised. Privately owned recreational boats tend to use gasoline as fuel, the flash point of which is well below room temperature. This is undoubtedly the principal fire hazard with such vessels. Gasoline vapour is much denser than air and on leakage will not disperse rapidly to levels below the lower flammability limit as, by reason of its being less dense than air, methane will.

Ibadan tanker truck explosion

Occurring in November 2000 close to the Nigerian city of Ibadan, this accident claimed between 100 and 200 lives; the precise number is not known. A tanker bearing gasoline was travelling at high speed when it encountered a queue of stationary cars. Its brakes being ineffective, it failed to stop and ran into the queue of vehicles. The tanker overturned and its payload burnt as a fireball. Many drivers of the cars in the traffic queue which the tanker had entered were burnt to death, whilst others were killed by flying debris from the fireball which had been accompanied by a significant overpressure. One of the most disturbing aspects of the matter is the suggestion that the queue of traffic had developed in the first place because police were stopping cars and demanding protection money from their drivers.

Ilyushin IL-76

Air tanker manufactured in the former Soviet Union having a payload of 11 000 gallons. This is over three-and-a-half times that of many "water bombers" currently in use in North America, including the **P3 Orion** the **C-130 Hercules**. It is about the same as the capacity of **Tanker 910** which, of course, is the only one of its kind, whereas there are many IL-76s in service. They are not currently allowed for use in the US and this was a point of some controversy during the October 2007 forest fires in California.

Inchon

South Korean city, the scene in 1999 of a karaoke fire in which 55 persons died and many more were injured. Ignition of paint thinner is believed to have been the origin of the fire. There were no sprinklers or fire alarms, and too few exits. Comparable in terms of loss of life was the karaoke fire at **Palembang**.

Indian Hills, KY

This municipality passed in 2007 an ordinance requiring that all new residential properties be fitted with fire sprinklers. It was the first "community" in Kentucky so to do, and invoked the following argument in promoting the ordinance. Several municipalities in greater Chicago are requiring new homes to have sprinkler protection, yet the climate there is measurably colder than in KY and fire risks correspondingly lower.

Inergen

Extinguishing agent composed of three gases: nitrogen, argon and carbon dioxide. It acts similarly to **Argonite**, with which it shares the advantage of total absence of water, though it is slightly denser. It is seen as a replacement for Halon 1301.

Inerting

The hazards due to the vapour/air mixture above the surface of a stored flammable liquid can be eliminated by substitution of an inert gas for the oxygen in the space sufficiently to make the gas/vapour mixture there too rich to ignite. One way of bringing this about is to pressurise the space with inert gas and then vent back to atmospheric pressure, an operation which will sometimes have to be carried out several times before the target oxygen concentration is reached. An alternative is **sweep purging**. (*See also* **Aircraft fuel tanks, fire hazards with**.)

In-flight electrical fires, number of diversions due to

Internationally, about three flights a day are so diverted. A large passenger aircraft has of the order of 90 miles of wire. Factors leading to malfunction and heating include contact of wiring with dust and other debris and insulation failure.

Integral design of fire appliances

A present-day fire appliance is a vastly complex piece of machinery and one might expect it to be in some degree "assembled" rather than "manufactured" in that components will have been received from many sources. Integral design principles are followed by many producers of fire appliances, primarily because it is better engineering practice but also because it prevents possible conflicts if a warranty issue arises. For example, if the stabiliser jacks, which prevent an appliance carrying aerial ladders from tipping when the ladders are in use, are unsatisfactory, one might reasonably take up the matter with the manufacturers of the stabiliser jacks. They in turn might argue that the position of the centre of gravity of the appliance was the errant factor, not the stabiliser jacks *per se*, and that the company which produced the appliance in its final assembled state are liable under the warranty. Integral design helps to prevent this sort of conflict.

Internal floating roof

Device frequently fitted to storage tanks for flammable liquids, including crude oil. It contacts the surface of the liquid and in so doing eliminates vapour which, over time, can build up to flammable proportions. In such a tank there will also be a fixed roof, and the volume between the upper part of the floating roof and the inner roof, which varies as the liquid surface goes up or down, is filled with air at atmospheric pressure.

Intumastic

Sealing compound for use in building, manufactured by Firetherm in the UK. It displays **intumescence** and has in recognised tests been shown to afford four hours of fire resistance. Other Firetherm products include intumescent paints.

Intumescence

Phenomenon whereby a solid substance on heating swells and then hardens into a char-like material. It is extensively applied to fire protection. For example, if the edges of a fire door are treated with an intumescent material, its action will cause the door to seal and thereby ensure its role in preventing fire spread. This will of course be at a temperature well above that at which human egress is possible so the role of the door in such egress is not negated. An intumescent material will not be a single chemical substance but a composite, for example sodium silicate incorporated into a plastic such as PVC. Such a composite can be made into strips to place on the edge of a door. Terminology for such a composite is that the plastic is the "carrier" and the additive material the "core". Graphite finds wide application as a "core" component. There are also intumescent materials which are supplied as lacquers for application to wood surfaces. These are used for fittings in retail outlets where a user will require a good quality of finish on the wood surface as well as effectiveness of fire protection. In offshore oil and gas production **Chartek 7** is currently the most widely specified intumescent fire retardant.

Iodotrifluoromethane

Chemical formula CF_3I, used as a fire extinguishing agent as a substitute for **Halon 1301** for air and naval applications in particular. It can be used in fighter aircraft to prevent ignition when the fuel tank of an aircraft is under threat. Halon 1301 was previously widely used for this purpose. The USAF came under pressure to develop a substitute for Halon 1301 which would not harm the ozone layer, and iodotrifluoromethane was the result.

IR-800/1 spark detector

A product of Sense-WARE in the US, this works by impingement of thermal radiation in the wavelength range 0.7–1 µm on to a semiconductor and needs mounting at a position free of solar or artificial light. It is particularly suitable for use in ducts. Its scope of "vision" is 110°, and it is common for two to be placed in a duct in complementary orientations.

IRcel®CH$_4$

Recently introduced detector for methane, from City Technology in the UK. It works by infrared. The IR source is a heated tungsten filament.

Island View

Hydrocarbon storage site in Durban South Africa, the scene in September 2007 of an explosion resulting in one fatality. It is believed that a road tanker at the depot caught fire and that that was the cause. Subsequent escalation involved eight stationary tanks at the site. Over 28 hours after the explosion, leaked hydrocarbon from the tanks was present in concentrations of several parts per million in nearby areas.

Isle of Wight Airport, fire at

On New Year's Eve 2007 there was a fire at the IoW airport, which badly damaged a restaurant. Adding to the loss is the fact that the restaurant contained many 1930s air travel memorabilia. Just over a year earlier (October 2006) there had been a major fire in a hangar at the same airport, known locally as Sandown Airport. The Isle of Wight Fire and Rescue Service, the majority of whose fire fighters are hired on a "retainer" basis, attended each fire.

ISO, and CAF systems

The Geneva-based body ISO issues documents relevant to fire protection, which are invoked in legislation, in litigation and in fire insurance. **Compressed air foam** systems, though seen by some experts as having the potential if not to revolutionise certainly strongly to influence routine fire fighting, have not at the present time featured in ISO standards. Given the influence of ISO and the ramifications of its standards relevant to fire fighting, this could be seen as a difficulty in the proliferation of **CAF** systems and installations. Interestingly, the Texas legislature has taken a lead in approving an addendum to ISO procedures for assessing fire insurance premiums in situations where CAF is available. Under this addendum a suitable CAF system will qualify a building for "insurance credits". An identical building with an equivalent CAF system in another state of the US would not obtain these credits.

Isodri®

Range of composite materials for fire fighter apparel developed for moisture resistance in particular. They use widely available substances including **Crosstech®**, **Nomex®** and **PBI™** in a layered structure, the outer finish being Teflon.

Istanbul Airport, fire in 2006

Fire fighting aircraft normally used for forest fires were brought into service for this fire at Istanbul's Ataturk Airport. Fifty conventional fire appliances also attended the fire, in which there were no serious injuries although a small number of persons were treated for minor injury or for smoke inhalation. Three buildings at the airport were destroyed. Initial views were that the fire was electrical in origin.

Numerical Exercises on Heat Transfer

Dr. J.C. Jones, Department of Engineering, University of Aberdeen, UK

A companion text to *Hydrocarbon Process Safety* containing 90 detailed numerical problems in heat transfer which vary in difficulty – a valuable teaching/learning tool.

> ...the book is a very valuable contribution to both engineering students and practising engineers. Indeed ...others who work in the industrial area will find the book to be of value ... *Fuel*

...the book is deemed to provide a useful text for students in the application of basic heat transfer principles to the safety of hydrocarbon operations. ...will find use within tutorial sessions... ...it will complement its companion book, *Hydrocarbon Process Safety*... *Process Safety and Environmental Protection*

ISBN 978-1904445-13-5   242 × 175 mm   102pp   diagrams   softback   £9.99

J

Jacksonville, FL, fire in 1901

Downtown Jacksonville was destroyed by a fire in May 1901 which began when hot ash having exited a chimney landed on an ignitable material. Ten thousand persons became homeless and the death toll was seven. The rebuilding was according to styles then being introduced by Frank Lloyd Wright, although he personally did not participate. (*See also* **Hotel Roosevelt fire**.)

Jacobus, PA

The **Good Will Fire Company # 1** of this town has appliances which include an engine initially manufactured by American LaFrance for the FDNY, entering service in 1981. After 16 years of service it was acquired by the Good Will Fire Company in Jacobus and rebuilt. It remains in service at the time of going to press, having a 1250 gallon tank and a pumping capacity of 1250 gpm. It sports a striking white and pale green livery. This appliance is however now relegated to second place, there having been purchase of a more recent American LaFrance engine in 2006.

Jamaica, recent acquisition of fire trucks from Japan

Jamaica has few assets and an uncertain economy. Such a country benefits from help by a highly industrialised one such as Japan and in that spirit two second-hand fire trucks in good condition having seen service in Japan were gifted to the Jamaica Fire Brigade recently. There is, however, still some way to go before Jamaica's fire fighting facilities are of a high standard. These include recruitment and professional training of fire officers and rebuilding of the central fire station in the capital Kingston. Japan has also recently donated a fire truck to Thailand for use at **Dokkhamtai**.

Japan, some accidental fires in recent years

These include one at a refinery in Hokkaido in September 2003, where an earthquake had destabilised a storage tank of crude oil. At the same refinery two days later a quantity of naphtha leaked and ignited. The result was a fire involving much of the surrounding area which necessitated the mobilisation of many fire crews. Also in 2003 was the fire at the ExxonMobil facility on Nagoya. Export of the then new Toyota Prius hybrid car in late 2005 was delayed by a fire at a Toyota factory in Japan. The fire started in the part of the factory where the body shells dry after painting. Moving from industrial fires to domestic, in June 2007 authorities in Japan reported that 54 fires had occurred during use of electric ovens manufactured by Matsushita. This manufacturing company, whose brand names include Panasonic, is a major producer of such goods having opened for business in 1918. The recent difficulty with the ovens was traced to the controller and purchasers have been offered free repairs.

Refuse-derived fuel (RDF) is obtained from municipal waste from households, possibly supplemented by industrial and trade waste, having been dried, shredded and possibly pelletised. In Japan in the early years of the 21st century there were a number of fires and explosions in stored RDF, one of which killed two fire fighters. The cellulose component of RDF decomposes at temperatures as low as 100°C to give flammable gas and vapour products, and this is believed to have happened during incipient self-heating of RDF, resulting in the explosive behaviour.

The spontaneous heating of rags soaked in oil is well documented. A bin containing rags contaminated, for example, with linseed oil can ignite through self-heating as the oil has the effect of promoting oxidation of the fabric from which the rag is made. In Nagoya in 2005 there was a fire believed to have been initiated by a fabric having been in contact with biodiesel. At **Takarazuka** in 2007 there was a fatal fire in a karaoke bar. (*See also* **Mah-jongg parlor, fire at; Mazda Plant fire, 2004; Myojo 56 Building**.)

Japan, systematic registering of fire appliances in

The length of reign of the Emperor of Japan in the year the appliance was first brought into use is incorporated into the registration mark displayed on a fire truck. The present Emperor Akihito has been in the role since 1989, hence a fire appliance registered in that year would have a "1", whereas one registered in 2009 would have a "21". This is preceded by an alpha character denoting the type of appliance, e.g. P for pumper, and is followed by a number signifying sequence, "*n*", for the *n*th appliance of that type in the fleet. So a pumper first registered in 2000 being the 20th acquired would be P 12 20.

Japanese tree lilac

Botanical name *Syringa reticulata*. Tree species, used as a **firewise plant** in parts of North America, including Alaska. It grows more slowly than many such trees.

Jay, FL

Location of a prison camp where in 1967 there was a fire which killed 37 inmates. At that time convict labour was used in Florida and at Jay prisoners detailed to work on the roads were housed in former Second World War barracks which were known not to be safe with regard to fire. The camp had, until shortly before the fire, been occupied by black prisoners only and when 15 white inmates were taken there to "integrate" it, confrontations resulted. One such escalated into a riot during which the building was intentionally set on fire by one or more of the prisoners.

Jersey City, NJ

The scene in October 2007 of a fire in a high-rise residential building still being constructed. It occurred close to the top of the 18-storey structure and it appears that water for fire fighting was not available above the fourteenth storey. Wood for use in internal fittings was being stored near the place where the fire began and this ignited. The building support was not threatened and construction will resume after repairs to the areas affected by the fire.

Jet-X®

High expansion foam concentrate having hydrocarbon surfactants, for use not only with fresh water but also with sea water. With a suitable proportioner and foam generator it can achieve a volume ratio of up to 1000 although such large ratios are not necessarily required and lower ones might be aimed for by adjustment of proportioner and generator conditions. Alternatively, protein-based foams are available with maximum ratios of 100 or less. Jet-X® is sometimes applied without subsequent foam generation to wood fires, in which case its function is that of a wetting agent.

Jewel 6L

"Wet chemical" fire extinguisher utilising potassium acetate and potassium citrate, suitable for Classes A and F of fire. Of 6 litre capacity, it will discharge fully in just under a minute and will "throw" to 2 to 3 m. Sister products include the Jewel 6L **AFFF** extinguisher, recommended for Classes A and B, and the 1 kg ABC Jewel Dry Powder extinguisher which uses **monoammonium phosphate**.

JFK Airport, bird strike at in 1975

A DC-10 aircraft taking off from New York's JFK Airport in November 1975 drew several gulls into one engine. The pilot returned to ground, and the aircraft ran off the runway and was destroyed by fire, which had begun in the affected engine and was exacerbated by overheated brakes. The death toll was 138.

Jharia

Location within India, the scene of coal fields. Fires in coal mines are common and sometimes continue burning for many years. The two complementary factors are spontaneous heating of the coal itself and the presence of coal bed methane. The first detection of a coal mine fire in the Jharia fields was in 1916 and at the present time there are about 70 fires in the Jharia coal fields. An additional concern with such fires is the carbon dioxide they release. It must be remembered that the burning of the coal within the mines has been to no useful purpose whatsoever, so the carbon dioxide it produces does not come within the scope of carbon accounting and is simply an excess.

Joelma Building

Building in São Paulo, Brazil, which in February 1974 was the scene of fire caused by an electrical fault in the air conditioning system. It appears that the inside of the building was cluttered with combustible items including paper, and that the internal walls were made of wood. The fire began on the twelfth storey and persons naturally tried to escape by ascent to the roof. Had there been a heliport on the roof, the consequences would have been less severe. In the event the death toll was 188.

Joy, IL

Scene of recent delivery of a new pumper built by **Alexis**. It is 9 m in length and is powered by a 330 hp Cummins diesel engine. Its Hale pump can supply 1000 gpm and the water tank capacity is 1000 gallons. Of comparable size is the pumper recently supplied by Alexis to Freeburg, IL. Of length 9.3 m, this has a Caterpillar 400 hp engine, a 1250 gpm Hale pump and a 1000 gallon water tank. It also has a foam proportioner and electrical power generation.

K

K1000 Elite

Thermal imaging camera for use in fire fighting, responding to infrared radiation in the wavelength band 8–14 µm. It has a rechargeable battery which will enable the camera to operate for three hours between rechargings.

Kansai Airport, diversion of an Australia aircraft to in 2005

In August 2005 an Airbus A330 aircraft registered to Australia and operated by an Australian airline took off from Narita Airport in Tokyo bound for Perth, Western Australia. It diverted to Kansai Airport, 50 km from Osaka, after smoke was found to be coming from the cargo hold. By the time the smoke was discovered, the aircraft was at cruising height. After the landing at Kansai there was evacuation during which one passenger sustained injuries classified by the Aircraft and Railway Accidents Investigation Commission (ARAIC) of Japan in their report of the incident as serious. Eight other passengers received injuries classified as minor.

Kansas, forest fire protection in

For 40 years fire services in Kansas have benefited from the **Federal Excess Property Program**, obtaining fire trucks, generators, water pumps and illumination from that source. The fire trucks were usually received as general-purpose trucks and retrofitted for fire use, this being typical of how rural fire departments in the US often have to take a "make-do-and-mend" approach. On the positive side, spare parts as needed can be obtained at below-market prices through another Federal program. Usually amongst a volunteer fire fighting crew there will be an electrician, a mechanic and so on who can keep the vehicles maintained. The trucks on loan to Kansas through the Federal Excess Property Program would cost US$45 million to replace with new or refurbished trucks if the Program ceased and the trucks on loan had to be returned.

Kaprun mountain railway accident, November 2000

One hundred and fifty-five people died when a train for conveying passengers up and down a mountain caught fire at Kaprun, Austria on the very first day of that year's skiing season. The train was inside a tunnel at the time of the fire. Afterwards there were claims that doors from the train had not been openable and that both sprinklers and fire extinguishers were absent. Suggested causes included flammable liquids or fireworks taken on the train, and lubricant from the train having overheated and spilt on to the track subsequently to ignite. The most recent safety check on the rail system at Kaprun had been in 1997. There had been a rise in the number of holidaymakers at Kaprun, which had not been accompanied by an increase in mountain rail capacity and at the time of the fire the train was overfull. Sixteen officials of the company which ran the train were charged with criminal negligence, and all were acquitted.

Kapton®

Material developed by DuPont, a polyimide stable across an extremely wide temperature range. Its applications are wide and include spacesuits. In fire protection, it finds use in the manufacture of **SCBA** and of **escape hoods**. Once widely used as insulation for electrical wiring in aircraft, it finds such application now only as composites including KT , **KKF** and **TKT**. Kapton® alone for such a purpose is avoided because of its tendency to crack and the possibility of loss of insulation that that entails. (See also **Boeing 737, two incidents with**.)

Kettle Foods

This company's plant near Norwich, UK was the scene in early 2005 of a blaze having begun in a vat of oil, which was attended by over 80 fire fighters. Other relatively recent fires in the food industry include that at the **Moy Park Chicken Factory, Co. Down** and that at **Premier Foods, Bury St. Edmunds**.

Kidde 0915Uk

Smoke alarm, working by ionisation. The sounder it activates has an output of 85 dB. There tends to be a difference in response times between ionisation smoke detectors and photoelectric ones. Kidde themselves recommend a "belt and braces" approach by installation of both.

Kidderpore

City in West Bengal. When an initially fairly minor fire occurred there in November 2007, fire appliances took an hour to arrive and many slum homes nearby were threatened. Occupants of the homes rioted and police intervention was needed before the fire fighters, once they did arrive, could go about their duties.

Kimberly Clark factory fire, 2004

This fire at the Kimberly Clark factory at Northfleet, in northern Kent close to the Thames estuary, was one of the largest industrial fires to have occurred in that part of England for some years and is believed to have begun with a malfunctioning piece of machinery. Only about three years later there was a less serious fire at the same factory.

King Cobra™

Trade name of a **fire fighter hood** made of a blend of **Nomex®** and **Lenzing FR®**. It is manufactured by PGI whose HQ is in Minnesota. The lining, made of a proprietary material, has the feature of removal of perspiration by capillarity.

Kingston, OH

The scene in December 2006 of a fire during the filling with gasoline of a pickup truck. Static electricity is believed to have been the initiating factor. Sparks were first observed when the nozzle from the petrol pump was removed from the vehicle. Previously, continuity between the nozzle and the vessel's bodywork had enabled charge to be removed to the avoidance of spark formation.

KKF

Composite substance for electrical wiring in aircraft, consisting of two layers of **Kapton®** with fluorinated ethylene propylene (FEP) as the outermost layer. It is used in the British Aerospace 146.

Knobcone pine

Botanical name *Pinus attenuata*, tree occurring on the west side of the American landmass from southern Oregon to Baja California and most prevalent in northern California. It is classified as a **highly flammable tree**.

Koilosava

Village in the Uttar-Pradesh region of India, scene of a major fire in March 2005. The fire began during cooking in a wooden hut and rapidly led to ignition of 40 neighbouring huts. Occupants of the affected huts lost all their possessions and cash aid given by the government was so scanty as to arouse indignation and a campaign for more adequate help.

Kronenburg

Dutch manufacturer of fire fighting vehicles, now part of Rosenbauer. The airports of London use between them many Kronenburg **ARFF** vehicles. Heathrow airport has ten or more; Gatwick has half a dozen, Luton Airport three and Stansted five. Many of these entered service in the mid 1990s, before Rosenbauer acquired the company in 2000. Municipal fire trucks from Kronenburg are often on a Daf chassis and might be badged as Dafs.

Kronitex®

Family of fire-retardant substances developed and manufactured by Great Lakes, each composed of phosphate esters. The organic moiety of Kronitex® TCP is derived from cresols, that of Kronitex® TXP from xyleneols and that of Kronitex® CDP from cresols plus phenol. Across the range Kronitex® products are used in applications as diverse as power cables, mattresses, adhesives and resins. Each has been used to fire protect PVC.

KTH-50X

Series of portable pumps from Koshin, Japan's biggest pump manufacturer. A big plus for the range is their low weight: even the largest weighs only 77 kg. The pumps are of centrifugal design and use high-chrome iron as the impeller material. The top-of-the-range can deliver over 400 gpm. Honda engines are commonly used.

Kuwaiti oil fires, losses due to

During the Gulf War in the early 1990s many oil wells in Kuwait were torched, and they remained burning for days. The loss of oil is believed to have been about 1 billion barrels. At that time a barrel of oil was worth about US$30. There were also of course indirect losses and severe envi-ronmental consequences.

Kynar®

Halogen-containing polymer having a good degree of heat resistance finding wide application. It was once used as an insulator for electrical wiring in aircraft and featured, for example, in early DC-9s. (*See also* **Zotek® F**.)

L

Labadie, MO

The scene in June 2002 of a fire at a transformer, in which an estimated 1000 gallons of **transformer oil** burnt. Combined losses from fire damage and interruption to operations were about US$5 million. (*See also* **Brunsbuttel**.)

Ladder bucket

A.k.a. a *platform*, a partly enclosed structure at the end of an aerial ladder usually capable of holding up to four fire fighters who will operate **monitors** from it and use it as an entry point to buildings at high level. A ladder bucket is often made from aluminium, like the ladder itself, and has gates for fire fighter exit and entry. It will often have lights which fire fighters can use to "find their way". Some builders of fire appliances, including the **Morita Corporation**, call a ladder bucket a "cage". In November 2007 in Bronx, NYC a fire fighter received serious injuries when he fell out of a ladder bucket.

Ladder fuels

In a fire at a wildfire/urban interface a fire beginning with ground level burning might ignite a shrub, from which the combustion might propagate to a tree next to it and from there to a higher tree in a ladder effect. In fire protection of homes this can be avoided by eliminating such a ladder configuration from the plant layout. Guidelines apply, for example a minimum distance of nine feet between a shrub and a pine tree of height three feet.

La Scala

Full name Teatro alla Scala, opera house in Milan which re-opened in December 2004 after an extensive overhaul, which included upgrading of fire protection. Parts of the building particularly vulnerable to fire include a wooden gallery and the archives room. A **Hi-Fog®** water mist system was installed, the first time Hi-Fog® had found application to such a building in Italy. There are now over 300 Hi-Fog® nozzles in the building.

La Fenice ("The Phoenix")

This famous opera house in Venice is appropriately named in that since its opening, under that name, in 1792 it has twice been burnt down and rebuilt. It was so named partly because the building it replaced, the San Benedetto Theatre, had itself been destroyed by fire in 1774. La Fenice was destroyed by fire in 1836, and rebuilding took exactly a year. This building remained in use until 1996, when it too was destroyed by fire. The follow-up resulted in two convictions for arson, a custodial sentence being handed down in each case. In 2003 the present La Fenice was opened.

LA Fireboat 2

Fireboat operated by the Los Angeles Fire Department, entering service in 2003. Its water release capacity places it in the **FiFi III** class, the pumps and monitors having been provided by **Skum**.

Laptops, possible fire hazards of

The fire hazard of these devices became very clear when at a business conference in Osaka, Japan a laptop caught fire whilst in use. The manufacturer of that particular laptop recalled four million others. There were by that time well documented cases of fires originating at laptops which had damaged property, including furniture, although there were no records of injury to persons. The ignition source in a laptop is the battery. The laptop's own circuitry will draw only a minuscule current with no significant heating effects, but if through malfunction the voltage between the poles of the battery is applied across a short length of conductor, heating will occur. Such an effect is most likely to be internal to the battery in which case there will be high thermal resistance to escape of the heat generated, increasing the likelihood of ignition.

"Last chance" breathing apparatus

Developed by a fire fighter in Connecticut and being evaluated by fire departments from a number of states, this device consists of a cylinder of oxygen which a fire fighter can draw on in the event that air supply from his **SCBA** stops. An important factor will be its size and portability having regard to the fact that a fire fighter's equipment constitutes a considerable load, additions to which might be at the expense of his comfort and ease of movement.

Laurier Palace Theatre, Montreal

The scene on Sunday 9 January 1927 of a fire claiming the lives of 77 children. At the time of the fire the theatre was occupied by about 800 children who had gone there to see a film. One of the two "safety exits" was locked and the other made less effective than it would otherwise have been have been by inward-opening doors. At a funeral for those of the dead who were from Roman Catholic families (about half) a local prelate (speaking in French) took the opportunity to admonish society for Sunday cinema and for admission of children to cinemas on any day of the week, expressing the view that cinema threatened the moral safety of children. The Roman Catholic authorities locally continued the campaign he had begun and the following year a law was passed restricting access to cinemas in Quebec to persons aged 16 and over. This remained in force until the early 1960s. A more positive result of the tragedy is that building regulations requiring fire exits to have outward-opening doors came into force. The fire was approximately contemporary with that at **Paisley, Scotland**.

LDS 19

Distillate material for use as pump oil, having a flash point of 238°C, manufactured by LDS in Florida. From the same company comes LDS 234. This has a lower viscosity than LDS 19 and a lower flash point by about 20°C. A positive correlation between flash point and viscosity is intuitively expected for substances obtained from crude oil.

Lead plane

Whereas a helicopter in fire fighting duty can enter "stationary hover" mode in order accurately to drop water from the air, a fixed-wing aircraft cannot. Advantageous use of water dropped from a fixed-wing aircraft therefore requires a planned trajectory which the lead plane can mark out in advance. The Beech Baron aircraft has been so used, leading an **air tanker** to a fire and then tracing the trajectory.

Lead zirconate titanate

A.k.a. lead zirconium titanate or (most commonly) PZT, a ceramic material which does not occur naturally and was developed for its piezoelectric properties. It finds application to fire alarm systems. An alternative in such systems is quartz, which is of course a natural product.

Leather Seal™

Fire-retardant preparation for application to leather furnishings. It is available in a number of forms with differing proportions of active constituents:

these are sodium bromide, potassium bromide, boric and/or phosphoric acid. It can be applied at the tanning stage, or afterwards by spraying. Such treatment of leather can eliminate the need for a fire barrier material between the external leather parts of a piece of upholstery and a combustible material with which it is filled.

Leicester Paper Converters

The premises of this company, a paper factory, was the scene of a major fire early in 2008, which 50 fire fighters attended. The fire began in the roof, which almost a day after the fire was examined by **thermal imaging** for any possible sites of combustion redevelopment.

Lenzing FR®

Cellulose-based fire-resistant fibre material, in some ways comparable to cotton having undergone the **Proban®** treatment. It can be blended with **Nomex®** as in the **King Cobra** hood.

Liberty I

A compressor for filling air cylinders used in **SCBA** that can be carried in a fire truck such as a **rescue vehicle**. As an alternative a compressor can be supplied mounted on a trailer which a fire truck can tow. The Liberty I series from Scott, a major US manufacturer of such equipment, is an example. Supported by a single-axle trailer, the compressor can be powered either by electricity or by an installed diesel motor. It is supplied with two cylinders conforming to ASME specifications and optionally can be equipped with support for a **cascade system**, itself drawing on the compressor.

Lidcombe, NSW

This outer western suburb of Sydney was the scene in November 2006 of a major factory fire. Many homes were evacuated, and there were no injuries. It is known that cardboard and plastic packaging materials were present in large quantities at the factory, the walls of which collapsed as a result of the fire.

Liège, fire in home for the elderly

Hydraulic oil in the lift system is believed to have been the origin of this fire in 2007. Two lifts were destroyed, but local fire crews were able to extinguish the fire without evacuating the building.

Life Liner

Trade name of a **fire fighter hood**, manufactured by a company in Nova

Scotia and widely used. It is fabricated of **Nomex®**. Such an item, once having entered service, has a considerable life expectancy and long after it ceases to be in new condition its authenticity might need to be established in following up an incident: was a fire fighter who was killed or injured wearing a **Nomex®** hood or not? For that reason the **Nomex®** label is sewn on to the Life Liner hood in several places and is highly durable.

Lightning, electrical potential of

Often in the approximate range 10 to 300 MV (megavolt). Lightning is a common cause of forest fires in the US, in some states the most common.

Lima, Peru

The poorer countries have to take a "make-do-and-mend" approach to many of life's needs. Sadly, this includes fire fighting and in Lima (population seven million) the fire department is woefully under-capitalised, its appliances being very elderly and not properly maintained. In a fire at **Mesa Redonda** in 2007, much less serious than the fire at the same location six years earlier described elsewhere in this volume, the "casualties" were six fire appliances! Quite unfit for duty at the time they were called out to the fire, they failed in their pumping function during it. It appears from press reports that no funds for their repair would be made available, yet to take them to any subsequent fire would be a waste of fuel and effort. A light-duty fire appliance, called the Pride of Gloucestershire, has been donated jointly by the British Embassy in Lima and the Gloucestershire Fire and Rescue Service for use in parts of Lima having the most severe difficulties.

Limberg

French oil tanker which in October 2002 exploded whilst off the coast of Yemen. It was carrying a quantity approaching 400 000 barrels of crude oil at the time of the accident, which according to news reports began when the tanker was rammed by a small vessel. The tanker was double hulled and would therefore have been expected to be able to withstand some degree of impact without loss of payload. It was concluded from that that the vessel which collided with the tanker was travelling at high speed. Descriptions of the event emphasise explosion as well as fire. The vapour pressure of a crude oil depends on the type of crude, which hydrocarbon types are in preponderance, and the distribution of weight across the boiling range. There will also be dissolved gas (which, of course, is processed into liquefied petroleum gas, a.k.a. LPG, on refining). Explosions involving crude oil had been observed prior to the Limberg accident.

L'Innovation Department Store, Brussels

The scene in May 1967 of a fire in which 322 lives were lost, one of the worst department store fires ever. The building of five storeys was 80 years old at the time of the fire and floors and walls were made of wood. It is unclear whether the fire began in the children's clothing department or in a higher floor where containers of butane for camping were on sale. Malicious lighting of the fire at the store is not ruled out, as the store had an exhibition of American goods which had been the target of protest against American activity in Vietnam at that time. It is at any rate certain that on that day and previously fireworks had been lit inside the store by such protestors.

Liquid carbon dioxide

A carbon dioxide extinguishing system, intended to protect a building where fire has begun, will require much larger amounts than those in a single-use **carbon dioxide extinguisher,** which will be typically 2 to 5 kg. One or more containers of liquid carbon dioxide with piping and valves for reticulation will constitute such a system. Such systems, including one manufactured by **Nippon Dry Chemicals**, have also found marine application. Gaseous and liquid carbon dioxide systems are called respectively, for fire protection purposes, low-pressure and high-pressure carbon dioxide protection systems. A low-pressure one will be refrigerated and sealed enabling phase equilibrium to exist between liquid at about 255K and vapour at about 20 bar. Pressures in a gaseous carbon dioxide extinguisher, in which the gas is superheated, will be 10 to 20 times this. The cost of the need for refrigeration might be offset by the much thinner walls needed to contain the carbon dioxide in its low-pressure form. The capital investment in refrigeration would normally only be justifiable for a protection system, not for a single extinguisher. The exception is the **Mini-Bulk®**.

Liquid nitrogen, use of as a fire extinguishant

As yet exploratory, at Boston, MA and Sydney, Australia. This cryogenic material (boiling point 77K) has performed in a superior way to water in laboratory scale trials on the extinguishment of liquid pool fires of propanol and of diesel. In large-scale application it will have the advantage of being totally harmless once used in extinguishment, causing no threat at all to air or water.

Lithgow

This mining town in New South Wales on the western descent from the Blue Mountains was the scene in January 2008 of a transformer fire. About 200 litres of **transformer oil** are believed to have leaked. Fire fighters from two brigades attended and **SCBA** was worn. The transformer was new and had only recently been installed. In the same month there was a transformer oil fire at Port Allen, LA and one at Beaverton, OR. The latter resulted in a fireball.

Lithuania, ammonia fire in 1989

Ammonia is hazardous through its toxicity as well as through its flammability. Seven lives were lost in this accident, which resulted from failure of a refrigerated tank containing 7000 tonnes of ammonia. That which spilt near the tank burnt as a pool.

Local area discharge method

In the use of carbon dioxide for fire extinguishment, the technique of directing the carbon dioxide at particular installations to protect them in the event of fire and the absence of partitioning walls or separating floors. It contrasts with the **total area discharge method** where the protective effects of the carbon dioxide are aided by the restricting effects of walls, doors and/or floors.

Lockheed C-130K

One of the Hercules series, many C-130Ks belonging to the Royal Air Force have recently seen service in Iraq and in Afghanistan. There are proposals, not at the time of writing authorised, to get them retrofitted with fuel tank **inerting**, notwithstanding the fact that the C-130Ks still in service are by now very elderly. When a Hercules was brought down by ground fire in Iraq in 2007 all on board perished, and it is believed that the effects would have been much less serious if inerting had been in place.

Lockheed Electra

Long obsolete as an airliner, the Electra, having been suitably adapted, is now having considerable success as an **air tanker**. One, based in Red Deer, Alberta, entered service as recently as the 2007 fire season, having undergone the conversion to an air tanker in the US. Electra air tankers have a payload of about 3000 gallons of water and are therefore comparable in those terms to the **P3 Orion** the **C-130 Hercules** and the **Beriev Be-200**.

Lok-Block

Trade name for material containing **gypsum**. Doors with a fire rating of three hours are made from it.

London, fire deaths per year

Over the seven-year period 1999 to 2006, the deaths per year due to fire in London had a mean value of 48.6. By way of comparison, there were 104 such deaths in New York City in 2005 and 137 in Tokyo in the same year. We define an index ϕ:

ϕ = annual number of fire deaths/population to the nearest million.

The ϕ values for the cities are London 6.9, Tokyo 11.4 and New York City 13.0, giving London a clear edge over the other two cities considered. Further examples of such statistics are in the table below.

City	Period	ϕ
Chicago, USA	2004	11.7
Jakarta, Indonesia	1998	7.5
Manchester, UK	2005	5.3
Manila, the Philippines	2006	33.9
Taipei, Taiwan	2004	3.8
Toronto, Canada	2005	5.6
Singapore	2004	14.0

(See also **USA, fire death rate.**)

London Fire Brigade, fire fighter training

It is important to appreciate that a fire fighter is not registered in the sense that a nurse or a paramedic is. Although fire fighter training needs to be "portable", enabling a fire fighter trained in one brigade to move to another if he or she wishes, there is not the uniformity of training that there would have to be if a national or regional enrolment of fire fighters took place. The London Fire Brigade requires its novices to attend a four-month non-residential course at its training centre in Southwark, across the Thames from the City. During this time skills with ladders, hose and **SCBA** are acquired. First Aid is also taught. Completion of the course is followed by a posting at a fire station as a probationary fire fighter. All being well, recognition as a full fire fighter follows after one year. Most importantly, in any fire service training is on-going and further courses will be attended by a fire fighter as his or her career develops.

London Olympics site, fire at

A fire occurred during the construction of the site in Stratford, London, which is being prepared for the 2012 Olympics. The fire was in one of the existing buildings at the site which was being demolished and is believed to have started when insulating material was ignited by an oxyacetylene torch.

London Round Thread (LRT)

Standard fitting for an underground fire hydrant for attachment of a **standpipe** at the outlet. It has diameter of 2.5 inches. The material of fabrication is not part of the specification and LRT outlets come in a variety of materials, including nylon.

London schools, (absence of) sprinkler systems

In the building and ongoing maintenance of schools in the UK fire precautions are of a high standard and are primarily focused on the number of exits and the speed at which children can be taken to safety via those exits in a fire. This is implemented through such issues as the floor area of the main hall and (even more) of a room in which there is an increased risk, such as a chemistry laboratory or a carpentry workshop. The case for sprinklers might not be seen as a very pressing one were it not for arson, there being over 200 intentionally lit fires in London schools each year. The London Fire Brigade in 2006 produced figures to the effect that fewer than 10 of the 2474 schools under local authority control in London were at that time protected by sprinkers. There is consequently a campaign currently to have the schools fitted with sprinklers, and insurance companies advise that the cost of fitting a school building could be recovered as reduced insurance premiums in about eight years. There is a more humanitarian side to this: for children to lose their books and portfolios of work through destruction of a school by arson can be educationally damaging.

Lord Williams's School

A man was arrested on suspicion of arson after a fire at Lord William's School in Oxfordshire, England in mid-2007. It is believed that the fire services were called over an hour after the fire started. The fire was attended by 65 fire fighters.

Los Angeles Airport, tank fire at in 2006

On 1 June 2006 there was a fire involving stored jet fuel at LA Airport. The tank was far from full, holding in fact only about 1000 gallons of fuel. **AFFF** was used in the extinguishment (which took about half an hour) and cooling was achieved by directing water at the tank outside walls from **monitors**.

Los Angeles Fire Department, acquisition of AEDs by

It was announced in January 2008 that 425 **Automated External Defibrillators (AED)** would be bought for the LAFD. They will be supplied by leading manufacturer Cardiac Science whose HQ are in Bothell, WA. They hold over 80 patents appertaining to the AED.

Low-e glass

Low-emissivity glass. These reflect thermal radiation, keeping it on the same side of the glass as the fire. The alternative would be for the thermal radiation to be absorbed and conducted through to the other side of the glass, resulting in heating of the glass and eventual breakage and loss of fire protection function. Borosilicate glasses (including Pyrex™) naturally have fairly low emissivities. With other sorts of glass the emissivity can be lowered in one of two ways: by controlling the precise composition, including the iron content, or by use of a thin metallic oxide layer.[21] Another possible means of controlling radiation from glass surfaces is to make the glass as thermally bulky as is consistent with its role in the building and impact safety requirements. This means larger thicknesses and densities of glass, and high heat capacities. In this way the response of the glass to heat from a fire, be it radiative and/or convective, is sluggish. Whatever the emissivity, radiation from the glass depends upon the fourth power of the temperature, so even modest temperature control by this means leads to significant reductions in the radiative flux from the glass.

Low-sulphur diesel, electrostatic hazards with

When crude oil is refined, sulphur concentrates in the higher boiling fractions. Diesel has the highest boiling range of all the atmospheric distillates and so tends to be high in sulphur. The sulphur content can be reduced by hydrotreating fairly straightforwardly, but such treatment tends to reduce the electrical conductivity, thereby creating a static electricity hazard. Incorporation of **fuel conductivity improver** is in such circumstances necessitated. One such for use with diesel, called by its manufacturers SR 1795 (Antistatic), has amongst its ingredients dodecylbenzenesulphonic acid and an amine polymer.

Luminaire

Term for an emergency lighting installation, the lamps themselves and the electrical components. A luminaire may be maintained or non-maintained in the sense of those terms in emergency lighting.

Lynne Moran

Tug vessel operating close to Port Arthur, TX, having entered service in 2005.

---

21 Sometimes when low-e glass is fitted to buildings the motive is not fire safety but energy conservation. This is because the glass limits long-wavelength absorption by the glass from a warm room and so prevents conduction of heat from the room to the outside via the glass.

She is classified **FiFi I**, having two pumps for fire fighting each powered by a 900 hp diesel engine.

The Principles of Thermal Sciences

Dr. J.C. Jones, Senior Lecturer, University of Aberdeen, UK

A unified treatment of thermodynamics and heat transfer useful to chemistry students but also students in engineering who will learn how the classical and statistical approaches to the subject complement each other. Numerical examples make it an ideal text for private study or tutorial use.

> ...The author should be congratulated on a clear exposition of a difficult topic in a very well written book, which is well prepared with crystal clear text and line drawings. ...it is a good buy and can be thoroughly recommended as excellent value for money... *Fuel*

ISBN 978-1870325-18-9   234 × 156 mm   176pp   illustrated   softback   £16.95

M

M25

Motorway which encircles London. On 30 August 2005, a lorry carrying hydrogen peroxide exploded whilst travelling along the M25. The only person injured was the lorry driver. However, debris from the explosion necessitated closure of all lanes of the motorway in both directions and the truck itself was totally destroyed. Hydrogen peroxide decomposes according to:

$$H_2O_2 \rightarrow H_2O + \tfrac{1}{2} O_2$$

this reaction being accompanied by the release of 2.5 MJ per kg of the peroxide so reacted. The hydrogen peroxide explosion on the M25 scattered debris across a wide area and wrecked the vehicle carrying it, therefore there must have been a major overpressure.

M30 Tunnel

Part of an urban motorway in Madrid currently being built. Its active fire protection will include a **Hi-Fog®** system along parts of it. This will on activation release water at a rate capable of controlling a truck fire long enough for occupants of the tunnel to evacuate.

Macron 55 Series

Hose reels, available for 19 mm (0.75 inch) or 25 mm (1 inch) hose. The reel can be either manual or automatic; however, the manual form has an interlock device such that the hose cannot be unreeled until the water is switched on. Where long runs of hose are to be accommodated, the reel becomes wide, and if its protrusion into a room or corridor is a problem, it can be recessed into the wall. Cabinets are available from the reel manufacturer, which might simply be mounted on the wall. If the reel is recessed, the "cabinet" incorporates the space occupied by the reel over which a pair of doors is placed. The Macron 70 Series uses wider diameter reel so as to provide for longer runs of wire without the width problem. A hose cabinet is a piece of **detention fire equipment**.

Madrid: "A tale of two fires"

Two major fires in Madrid, separated in time by about three years, have been compared in the fire protection literature. The table below summarises the comparison.

	Year	No. of storeys	Whether sprinklered	Amount of extinguishing water/litre	Estimated insurance payout /€	Ultimate fate of the building
Fire 1	2002	7	Sprinklered	≈ 26 000 from sprinklers	1.75 million	Back in use the following day
Fire 2	2005	32	Not sprinklered	> 6 million, from fire fighting hose	300 million	Total loss

Both fires occurred out of working time: the first on New Year's Day, which is a holiday, and the second late on a Saturday night. It is always important to compare like with like otherwise comparisons can mislead. We therefore note that the second fire occurred in a far taller building than did the first. Even so, evidence of the benefit of sprinklers is very convincing. (See *also* **M30 Tunnel**.)

Maersk M Type

One of relatively few vessel types having **FiFi III** classification, the Maersk M Type is used in the management of mobile offshore structures, for example in anchor handling. It owes its classification to four **monitors** each supplied with water by its own pump. Maersk M Type is an example of an anchor handling tug supply (AHTS) vessel. Such will more commonly have **FiFi I** classification and Maersk themselves manufacture AHTS vessels of this classification, largely for use in Canadian waters. There are also Maersk vessels in use in the UK sector of the North Sea.

Maersk R Type

This predates the **Maersk M Type** and there are three sub-types: the **R**ider (**FiFi III** rating) the **R**over (also **FiFi III** rating) and the **R**etriever (**FiFi II** rating).

Magar

Ship of the Indian Navy, the scene in February 2006 of an explosion in which three on board were killed and 19 others injured. The vessel was transporting redundant military explosive devices for disposal at sea. One such device went off during handling.

Magnesium hydroxide, use of as a fire retardant

At temperatures of incipient combustion, this reacts according to:

$$Mg(OH)_2 \rightarrow MgO + H_2O$$

and in that respect acts entirely analogously to aluminium trihydroxide (ATH) as a fire retardant. Preparations containing magnesium oxide are available for incorporating into elastomer products as smoke suppressants: one such is Humag®, developed by the UK elastomer manufacturer Hubron. Magnesium oxide is increasingly being used to fire protect **TPO** in place of halogenated retardants.

Mah-jongg parlor, fire at

Mah-jongg is a game played for financial stakes and a popular form of gambling in Japan. At a Mah-jongg venue in a sleazy part of Tokyo in September 2001 there was a fire which claimed over 40 lives. It was the worst fire in terms of loss of life in Tokyo for 19 years.

Maintained emergency lighting

Emergency lighting is required in order that persons can evacuate a building if the mains power is no longer working. Maintained emergency lighting is wired into the mains and is not reserved for emergencies, being part of the regular lighting system. However, it has back-up batteries which will enable it to function, typically for three hours, if the mains fails (or has to be switched off). The batteries in such an emergency lighting system are kept topped up by the mains.

Malibu, CA

The scene in November 2007 of a vegetation fire believed to have been accidentally started by a camp fire. Aided by strong winds, the fire advanced rapidly and many homes were destroyed.

Malila

Supply vessel for deep sea operations, entering service in 1999, currently registered to St Vincent and the Grenadines. It is classified **FiFi I**, having two **monitors** capable jointly of water release at 2400 m³ per hour and a "throw" of 120 m.

Malmaison Hotel

This hotel in Oxford, adapted from a former prison, was the scene of a fire believed to have been of electrical origin. Having been so adapted, the building was irregular in its internal structure, and because of this the local fire service had visited it some months earlier to note its configuration. This is believed to have been a factor in minimising damage from the fire.

Mammoth Lakes Yosemite Airport, CA

This took delivery of a new Colet Jaguar K/15S closely similar to that at Sonoma County Airport in September 2006. The airport accordingly raised its FAA index to B, signifying that it can receive aircraft up to 125 feet (38 m) in length. This requires an **ARFF** capable of supplying 1500 gpm. The newly acquired Jaguar can comfortably meet this criterion. The FAA Category so achieved corresponds (according to the present author's "translation") to about 5 in the **NFPA categories of airports**.

Mann Gulch, MT

The scene in 1949 of a forest fire in which 12 fire fighters, having been parachuted into the affected area from a **smokejumper**, lost their lives as well as a fire fighter who travelled to the scene "conventionally". Factors increasing the danger included rapid spread of the fire by wind and inexpert application of the **escape fire** technique. The fire, which became the subject both of a book and of a folksong, was followed by two ironic events. The foreman of the fire fighting crew W. Dodge himself survived but died from Hodgkin's disease five years later. A fire expert by the name of Gisborne from the US Forest Research Center personally led an examination of the site of the fire some months later in spite of having been diagnosed as having heart problems. He had a fatal heart attack on the last day of his visit.

Maria Ratsehitz Mission Hospice

During a fire at this building in KwaZulu-Natal, South Africa in 2007 the burning thatched roof collapsed on to a nurse who was helping with evacuation, with fatal effects. Three of the residents also died. The fire is believed to have been started by a cigarette.

Maricopa, AZ

The scene in 2007 of a fire at a tyre recycling plant. Fortuitously the wind took the toxic smoke in a direction away from dense housing. A state of emergency was declared, this enabling access to additional sources of water. Fire fighters from other localities, including Phoenix (20 miles away), attended the fire. It was the third fire at the plant; one of the two previous ones is believed to have been arson.

Massey Shaw

Fire fighting boat once in service on the River Thames, built in 1935 at a cost of £18 000. Her most notable performance was during a warehouse fire at Colonial Wharf, where the *Massey Shaw* provided a screen of water enabling the fire fighters on land access to areas which would otherwise have been impossible to enter.

Maxiforce™

Air lifting bag having found wide application in fire fighting. It is constructed of neoprene with **aramid fibre** reinforcement where some such bags use steel. A middle-of-the range Maxiforce™ bag inflated with air to 8 bar has a lifting capacity of 15 to 20 tonnes. Such a lifting bag was recently used by a team of fire fighters in Massachusetts who had been called to the scene of a plane crash. Fuel was leaking from the crashed plane, and the lifting bag was used to level the plane and stop the leak, enabling the remaining fuel to be removed and thus eliminating hazards due to the fuel.

Mazda BT-150

This 4WD vehicle readily lends itself to conversion to a light-duty fire appliance, there being space for hose, a water tank and a **vehicle-mounted pump**. Mazda have donated 35 such fire appliances to Greece as a result of the 2007 fires at places including **Peloponnese**. (*See also* **Ford F-550**.)

Mazda Plant fire, 2004

In December 2004 there was a major fire at the Mazda plant in Hiroshima, which began in the painting area. There was significant disruption to production, reflected in a 5.2% drop in share prices by a few days after the fire. (*See also* **Porsche factory explosion, 2008**.)

MBK Centre

A.k.a. Mahboonkrong, a shopping complex in Bangkok, Thailand housing about 2500 businesses. It is an eight-storey structure and its shell is largely composed of marble. Opened in 1985, it has ranked amongst the biggest shopping malls in Asia. On 21 April 2007 there was a fire which began at a Japanese restaurant on the seventh storey. Ventilation was, by criteria which would have applied in countries such as the UK, inadequate and toxic smoke built up the dangerous concentrations. The future of the centre is at the time of writing uncertain.

Although the King of Thailand is the longest serving monarch in the world today, having occupied his throne two years longer than Queen Elizabeth II has occupied hers, Thailand is not noted for its probity in public life or for its impartial maintenance of law. This has had many spin-offs in fire safety terms, for example at the Royal Jomtien Hotel fire in 1997 in which 91 people died. There is evidence that emergency exits from the hotel had been closed off with steel bars or with padlocks. Moreover, a fire appliance on its way to the hotel was involved in a road accident which claimed another 11 lives. In 1999, two years subsequent to the Royal Jomtien Hotel fire and in roughly the same part of Thailand, seven people died when hydrocarbon inventory at a refinery caught fire. In 1993 there was a fire in a doll factory in Bangkok in which 200 persons perished and many more were injured. In some non-urban parts of Thailand fire fighting facilities are dismally inadequate. The Government of Japan through its representation in Chiang Mai recently donated a fire truck with a pump to **Dokkhamtai**. (See also **Route 999 nightclub**.)

MD520N

Helicopter manufactured by McDonnell Douglas, suitable for fire fighting, in which case it is likely to have been fitted at manufacture with a **Bambi bucket**®. An MD520N so equipped is in service in Calgary.

Mecca, school fire in 2002

The city of Mecca, centre and focus of Islamic observance and activity, has a stable resident population of about 1.5 million. In 2002, 15 girls at a school in Mecca were killed in a fire. There have been allegations that the girls were prevented by the "religious police" (a.k.a. the Commission for the Promotion of Virtue and Prevention of Vice) from exiting the school because they were not wearing correct Islamic dress. This brought the religious police into conflict with fire officers whose protests were disregarded. Adding to the hazards was the fact that the school doors were locked. This is the usual practice in Saudi Arabia where separation of the sexes in schools is rigorously enforced.

Melamine

Heterocyclic organic compound, molecular formula $C_3N_6H_6$, widely used as a fire retardant for polymers in one of three forms. The first is pure melamine. The second is as a salt of an organic acid in which the melamine moiety is the positively charged part. An example of such a salt is that formed between melamine and cyanuric acid, called melamine cyanurate, with trade name melapur. The third is as an extended organic structure incorporating the melamine structure. The simplest of these, called melam, consists simply of one melamine structure attached to another at a nitrogen bridge. The formula given for pure melamine above signifies 67% nitrogen, and the ability to release nitrogen, thereby reducing the oxygen level, is one way in which melamine-based retardants work. Decomposition of melamine is endothermic and this also assists in retardation. Not itself metal containing, melamine acts like some retardants which do contain a metal in that it promotes char formation. There are also available fire retardants containing melamine plus an organophosphate, e.g. **Pyrovetax CP**. (*See also* **Basotect®**.)

Mercury™

An example of a portable **monitor**, that is one which will not be attached to a fire truck but will be set up *in situ* and once operating can be left unmanned. The capacity of the monitor in its basic form is 500 gallons (US) per minute $\equiv 115$ m^3 $hour^{-1}$. A version is available – the Mercury™1000 – in which this capacity is doubled. The inclination is fixed at 20° when the monitor is unmanned and can be manually varied up to 60° when manned.

Mesa Redonda

Location in **Lima, Peru** and the scene of a fire on 29 December 2001 which caused 291 deaths. At Mesa Redonda there is an outdoor market where, with a minimum of formal organisation and regulation, goods are bought and sold. Amongst goods so traded were fireworks, particularly saleable at that time of year, and when the fire began at about 6:00 pm there were large quantities of fireworks at the market. It appears that at one outlet for the sale of fireworks, the fireworks themselves had disintegrated and the pyrotechnic contents spilled out. Nearby buildings were affected by the fire and persons were trapped in them. Causes of death other than burns and smoke inhalation included electrocution: an electrical distribution facility was affected by the fire and this caused huge surges in supply with fatal consequences. Many lost their possessions and livelihoods, such as they were, as a result of the fire. Amongst the defendants when the case came to court was a senior police official who had authorised withdrawal of police from the area over the Christmas season. To this was attributed the fact that fireworks in a quantity exceeding the legal quota were at Mesa Redonda on the day of the fire.

Mesquite, TX

Suburb of Dallas, one of only about 50 communities in the whole of the US having the coveted class 1 on the **Public Protection Classification (PPC™)** scale, which it recently received. It was previously class 3, itself a very good result. Property owners in Mesquite could expect a drop in fire insurance costs of about 9% as a result of the transition from class 3 to class 1.

Metacaulk®1100

Example of a proprietary substance which provides passive fire protection by **intumescence**. It can be obtained in a liquid form resembling paint and applied where required manually or as a spray. Its chief use has been in the filling of gaps in building structures, which otherwise in the event of fire would provide a route for spread to a neighbouring room or passage.

Metal hydride actuator

Consider the generalised chemical reaction:

$$MH_n \leftrightarrows M + 0.5nH_2$$

where M denotes a metal. When a container of the metal hydride is heated, the released gas can activate a mechanical device. When the container is cooled, the reaction takes place in the reverse direction and the mechanical operation is also reversed because of the pressure drop. This is the basis of the metal hydride actuator. These find application to fire protection, for example in the operation of sprinkler valves.

Metal mesh, use of in explosion protection

In hydrocarbon/air mixtures explosion in the sense of thermal runaway is due to build-up of heat *and* of reactive intermediates. Contact of a mixture approaching explosion conditions with a metal surface will remove both heat and reactive intermediates, restricting in both ways the development of an explosion. Consequently metal mesh is sometimes installed in storage tanks and reactor vessels for hydrocarbons. There are aluminium alloys which have been developed for this purpose. Clearly, in the performance of such a device the metal surface area is important and this can be raised by going from beyond a simple mesh to an **anti-explosion pad**.

Metuchan, NJ Fire Department

This FD has been selected for mention partly on account of its operation of a heavy **rescue vehicle** built by Pierce in 1996. This has a 350 hp Detroit diesel engine and a power take-off (PTO) generator capable of 35 kW. It also has an illuminating mast which stands 35 feet high, the floodlight it supports being 1500 W. The Department has two Pierce pumpers and a

Pierce aerial. It also has a 1970 Mack Trucks pumper which is powered by a 276 hp *gasoline* engine, which will remain in use at least until 2009, possibly a little longer.

Metz

Company manufacturing fire fighting equipment having begun in Heidelberg, Germany and now headquartered in Pennsylvania. One of its major products is an aerial ladder. Metz aerial ladders have been incorporated into appliances made by **Angloco** amongst many other builders of such appliances.

Mexican orange

Botanical name *Choisya ternate*. Shrub, indigenous to Mexico and the US states of Texas and Arizona and a **firewise plant**.

Mi-26

The world's most powerful helicopter which, when used in fire fighting, can carry a container holding 5000 gallons of water. Six were made available by Russia during the 2007 forest fires at **Peloponnese**.

Microgrid

Flexible metal grid material which can be applied to an aircraft fuselage to protect it in the event of lightning strike. It is available either in aluminium or copper, each of which is an excellent conductor of electricity. The recommended level of application is one layer for the fuselage and two for the fuel tanks.

Middletown, MD

The volunteer fire fighting service in this town was itself the victim of arsonists in 2006, when a valuable **Seagrave** fire truck on loan to it was intentionally set on fire. It experienced major damage as did a vehicle belonging to one of the fire fighters which was parked nearby.

Midel 7131

Transformer oil containing synthetic esters, having a fire point of 322°C. Even when it does ignite it does so with much less smoke than a conventional transformer oil would produce on burning. These characteristics have earned Midel 7131 a recommendation from Factory Mutual.

Midship pump

A fairly imprecise or ambiguous term which nevertheless is encountered frequently in descriptions of fire apparatus in the US. When in 1996 there was a legal complaint against one supplier of pumps the definition used was: "truck mounted fire pump that meets NFPA 1901" making the term synonymous with **vehicle-mounted pump**. Usually the term means a pump mounted behind the cab of a fire appliance connected to the vehicle's own driveshaft. Its power is therefore potentially equivalent to that of the vehicle itself. A pump having its own gasoline or diesel engine might be mounted on the appliance in which case the term engine-driven pump applies. Since it is not being used as a portable pump, the weight restrictions applying to such are less important. The other possibility is use of a power take-off (PTO) where a hydraulic pump draws on the vehicle's own power but up to a lower maximum than that which a midship pump can draw. Both engine-powered pumps and PTO pumps tend to have lower water discharge rates than midship pumps.

Milford Haven, Wales

The scene in August 1983 of a refinery fire. The fire began in a tank containing crude oil. The tank had an **internal floating roof**, which some hours into the fire broke up and sank. Later **boilover** occurred and some of the oil exiting the tank as a result ascended to heights of hundreds of metres. It therefore spread widely on descent, and the fire which ensued occupied an estimated 1.5 hectares.

Mill Creek Volunteer Fire Company, DE

Operator of a **quint**, first manufactured in 1996 by E-One, a major manufacturer of such plant, based in Georgia. Its aerial ladders extend to 75 feet and it carries 500 feet of LDH hose (from the manufacturer of Pro-Lite). Its pump can deliver 2000 gpm.

Millgarth

Sister vessel of the **Anglegarth**, built one year later.[22] Both vessels are based at the Port of Milford Haven.

Milton, FL

To the satisfaction of the local fire department and to the benefit of residents, Milton, FL recently retained on review its rating of 4 on the **Public Protection Classification (PPC™)** scale. The correlation of merit of fire protection with position on the scale in not linear: 4 is a good result placing

---

22   For a striking photograph of the two vessels in tandem, each releasing water from its monitors go to http://www.ship-technology.com/projects/anglegarth1.html

a community having that rating in the top 10% within the US. In some quite populous states the majority of fire departments are class 9, which should be seen not as being close to the bottom of the scale in a negative sense but as a rating which the insurance industry will endorse. There is no such endorsement for class 10 communities, of which there are a good number. That means that no insurance premium bonuses can, in a class 10 area, be expected on the basis of "public protection". (*See also* **Selected US states, Insurance Services Office classifications for**.)

Mini-bulk®

Carbon dioxide for fire extinguishment contained in a cylinder in a low-pressure state, that is as **liquid carbon dioxide** in equilibrium with its vapour. The cylinder has a refrigerating enclosure which draws on the electrical mains. Unlike a "full size" liquid carbon dioxide system, the Mini-Bulk® is intended to be discharged completely once brought into use but, because of the higher density of the liquid content, will operate for much longer than a high-pressure extinguisher of the same internal volume.

Mission Hills, CA

The scene in December 2007 of the overturning of a truck/crane combination on a freeway. Los Angeles Fire Department attended the scene and their priority was to contain leaked hydraulic fluid and prevent it from igniting. The overturned vehicle contained 500 gallons of such fluid.

MLGS4-30W

Fire apparatus built by the **Morita Corporation**, having aerial ladders and a water pump each of the Corporation's own manufacture. This contrasts with North American practice where the pump will have been bought in, for example from **Waterous**. The pump fitted to the MLGS4-30W will deliver 800 gpm at a head pressure of 0.8 MPa. An example of the MLGS4-30W was recently taken to Shanghai for display at the 7th Shanghai International Fire and Security Technology Equipment Exhibition.

Mobil Aero HF

Hydraulic oil for use in aircraft composed of hydrocarbon base stock from crude oil with additional ingredients to adjust the viscosity. Its minimum flash point (closed cup) is 82°C. A sister product of similar composition is Mobil Aero HFA, minimum flash point (open cup) 93°C. Each has a moisture content not exceeding 100 ppm.

Mobile homes

Mobile homes, having been constructed at a workshop and towed or carried on a truck to where they are to be installed and occupied, are, in this discussion, distinguished from caravans for recreational use only. Details of very recent (all 2008) fires in the US in mobile homes so defined are given in the table below, which is followed by some comments.

Location	Consequences
Odessa, TX	Three deaths (all children) One adult very seriously injured
North Port, CA	One death
Jamestown, VA	Two deaths

Fire fighting authorities in various places have published a factor by which, according to their records, a fire in a mobile home is more likely to have fatal consequences than a fire in a conventional home. The present author has seen values for this factor ranging from 9 to 16. Even at the low end of the range, the fire danger of mobile homes is unequivocal. One reason is that space is likely to be at a premium in such a home; a fatal fire in a mobile home near London in 2002 was caused by too close proximity of an electrical heater to clothing. There is also the obvious point that flammable gas cylinders will often be kept in a mobile home, which will lack connection to the local natural gas supply. In Cornwall, England in 2005 a mobile home was destroyed through leakage of gas from cylinders and the two occupants were hospitalised.

Mobile phones, possible hazards of at filling stations

In *circa* 1999 the view that a mobile phone if switched on could provide an ignition source for gasoline vapour gained considerable currency. Accordingly many service stations required petrol purchasers to turn off mobile phones whilst operating a pump. For the radio frequency (RF) emissions from a phone to generate a spark is now known to be an impossibility because the power is so low (<1W). An RF transmitter for broadcasting might create an ignition source on encountering a metal loop with a gap, as once happened in Scotland, but here the power is many orders of magnitude higher than that of a mobile. There appears to be no confirmed case of ignition of vapour at a filling station attributable to a mobile phone.

Mobile tanks of hydrocarbon, electrical earthing of

When a road tanker bearing gasoline arrives at a filling station in order to replenish the tanks there, precautions against ignition by electrostatic discharge begin with the forecourt surface itself. The resistance of this is important in eliminating risk, and criteria apply. The conductivity can be raised by use of **antistatic paints**. Metal pipe work along the path of the fuel from mobile tanker to stationary one should be connected to earth. With a ship such earth is provided by the water with which the hull of the ship is in contact so anything in electrical continuity with the hull is itself earthed ("grounded"). This is usually seen as sufficient, although ship-to-shore earthing during the unloading of hydrocarbon liquids from tanker vessels is in fact mandatory at some terminals. When a vessel of hydrocarbon is carried by rail, it is grounded provided that there is electrical continuity between it and the metal wheels.

Mobil Hydraulic 10W

Hydraulic oil having a viscosity at 40°C of 37.7 centistokes and a flash point of 232°C. This flash point value is fairly typical for such a substance, being for example equal to that of one of the **Texaco Rando® HDZ** series. It is however much higher than that of **Mobil Aero HF**.

Moisture barrier

Fire fighting apparel needs to be resistant to water penetration, partly because enormous amounts of water are used in fire fighting. There therefore needs to be a layer of the fire fighter's "ensemble" which provides a moisture barrier. In fact such a barrier might endanger rather than help a fire fighter if it did not as well as keeping extinguishing and other water out allow perspiration to exit in the opposite direction from potential ingress of water. Special materials for the purpose have been developed and some endorsed by NFPA. These include **Crosstech®** and **RT700A®**. A further function of such a barrier is that it protects a fire fighter from infected blood during contact with victims.

Molybdenum trioxide

Chemical formula MoO_3, used as a flame retardant for PVC when, as a bonus, it also acts as a **smoke suppressant**. It has also been used as a fire retardant for wood. Molybdenum disulphide, MoS_2, is not itself a flame retardant but a "solid lubricant" and is frequently incorporated into such materials as nylon to improve their resistance to wear. As such it is sometimes an ingredient of sealants and adhesives to which fire standards apply.

Monel

Trade name for a series of alloys in which the dominant metal (typically 67%)

is nickel. There are also significant amounts of copper and of iron. It has a thermal conductivity, depending on the precise composition, of around $21\ Wm^{-1}K^{-1}$, which is at the high end of the range for stainless steels. Monel is a possible choice of material for the firewall of an engine nacelle.

Monica Wills House

Sheltered housing scheme in Bristol, England opened in October 2006. In December of the same year a fire occurred there, causing one fatality and major damage. The building had sprinklers installed and that the effects of the fire were not even more serious was attributed to that. However, the one part of the development where sprinklers were not installed was the enclosed car park at ground level and the fire is believed to have started there. A vehicle in the car park caught fire through an electrical fault and in what must have been flashover, or a process akin to it, the fire subsequently involved all 21 vehicles in the car park directly above which were residential parts of the building. Between the roof of the car park and the residences was foam insulation. There was abundant smoke, and smoke damage exceeded thermal damage; the total value of the damage was £5 million.

Monitor

Term used in marine or other fire fighting in particular to mean a device capable of releasing water for direction at a fire. The term is not synonymous in this context with "pump" and the number of pumps and the number of monitors need not be the same. This is true for example of fire fighting plant on the *Odin Viking*. One specification of a monitor is its "directability", the angle in a horizontal plane through which it can be moved. This is especially important for a monitor fixed on a fire truck, less so for a portable one which can of course be placed so as to face the fire. The **Vulcan RF™** has full 360° adjustability whilst the **Mercury™**, being portable, can be moved through only ± 20°. The inclination will also be adjustable, commonly up to about 60° from horizontal. In an **oscillating monitor** there is movement through an arc in the horizontal plane during operation. The term "water turret" is often used synonymously with "monitor".

Monnex

Dry fire extinguishing agent suitable for Class B and Class C fires, developed by the former ICI and synthesised by reacting potassium bicarbonate with urea. It is particularly suitable for fires involving polar liquids.

Monoammonium phosphate

Powder fire extinguishing agent. ABC20, ABC30, ABC40, ABC70 and ABC90 are **ABC powder extinguishers**, where the numerals denote the percentage of monoammonium phosphate. Monoammonium phosphate works by forming on heating a thin layer which restricts oxygen supply to the burning material. In so doing it leaves behind a sticky yellow deposit which, like the use of excessive water in fire fighting, causes "non-thermal damage". It is sometimes combined with **borax**. The compound finds application not only as a fire extinguishing agent but also as a fertiliser and world production for the two purposes is of the order of tens of millions of tonnes per year, in places where phosphate minerals occur, including Western Australia. There are enormous phosphate reserves in Saudi Arabia and there are proposals to use these to make phosphorus compounds, including monoammonium phosphate, for the market.

Monokote®

Fireproofing material manufactured by Grace Construction Products whose HQ are in Connecticut. It is based on **gypsum** and can be applied to concrete or to steel. It is available in a number of different forms for application by spraying or by injection. Monokote® is said to be the most widely used fireproofing material in the world. Recent "high-profile" uses include Canary Wharf in London.

Monsoon bucket

Flexible container which can be suspended from a helicopter in fire fighting duty and filled from a tank, river or other source of water whilst the helicopter is in flight. It has a valve at its base for release of water on to a fire and this is controllable by the pilot. An example is the Helifire™ series, developed in New Zealand, which range in capacity from 345 to 1100 litres. An accessory is the Helimix® foam proportioner, by means of which foam can be added to water taken into the monsoon bucket, the proportioning procedure being precisely controllable from inside the helicopter. Another well known range of monsoon bucket is the **Bambi bucket®**.

Monte Carlo Hotel, Las Vegas

A recent fire at this hotel is believed to have begun in one of the upper floors. The hotel was evacuated and residents found places in nearby hotels. That there were no deaths or injuries might well be at least in part attributable to the hotel being fitted with sprinklers.

Monterey, CA

The scene in 2007 of a studio fire which was to a large degree controlled by the sprinkler system. The building was over 70 years old and had had sprinklers installed in the 1980s. At about the same time a fire took place in a convalescent home near Hartford, CT which had been extinguished by sprinklers before the fire department arrived. This fire had begun with an electric fan.

Monterey pine

Botanical name *Pinus radiata*, an example of a so-called **highly flammable tree**. It occurs in northern California in places including Santa Cruz and also in parts of Baja California. It has been naturalised to many countries including Australia and New Zealand (where it is known as radiata pine).

Montgomery, PA, vehicle fire in 2004

A 20-year old Cadillac Eldorado was gutted by fire of electrical origin in Montgomery in January 2004. The car was unoccupied at the time and there were no deaths or injuries. As a case study in such fires it illustrates some interesting points. The car interior was destroyed: there was no damage at all within the engine compartment. The windows of the car were, however, blown out. This was undoubtedly due to build-up of flammable decomposition products from the upholstery and other fittings to a level in excess of the LFL. The overpressure needed to break the glass would have been due to the confinement. The owner of the Cadillac had, when he last left the car, detected by the sense of smell that something was burning. Warning of an electrical malfunction potentially leading to fire is provided by frequent blowing of fuses in the car's electrical system: they only blow if currents of an unsafe magnitude are occurring and such currents are of course a fire hazard. It is not recorded whether the owner of the Cadillac had recently had to replace a fuse.

Morita Corporation

Japanese manufacturer of fire appliances having been in the business since 1907 and now having a large production base and worldwide influence. Its products have been exported to other countries including the USA. Its range takes in all the types of appliance manufactured by its US counterparts such as Pierce Manufacturing, E-One and **Sutphen**. Morita appliances are sometimes doubly named to incorporate the name of the manufacturer of the chassis/cab base. There are for example two "aerials" badged as Mitsubishi/Morita currently in service in Bangkok. Isuzu fire vehicles frequently incorporate Morita components, including pumps.

Moscow, nursing home fire 2007

Sixty-two persons died in a fire at a nursing home in Moscow in March 2007. In terms of loss of life it was the most serious fire in post-perestroika

Russia. President Putin in addressing the nation on the tragedy made the point that the low standard of services such as fire fighting in Russia are all the more unacceptable in view of the high oil and gas revenues being received by Russia, which ought to make for prosperity.

Motor Sports Association, UK

One of the functions of this body is the promotion of effective safety measures for enthusiasts, for example fireproof clothing. The Standard which it applies to clothing is BS EN 533:1997, the sub-title of which is "Protective clothing. Protection against heat and flame. Limited flame spread materials and material assemblies". One of the many products it endorses on this basis is the Clubman race suit, manufactured by Advanced Wear and Safety (AWS). This is made from cotton treated by the **Proban®** process and can be obtained either "off the peg" or "made to measure". Having regard to the intrinsic element of flamboyance in motor sport, the Clubman and similar products are available in more than one colour, making for visual appeal as well as safety!

Motor vehicles, destruction of by arson

Because of improvements in design, accidental vehicle fires have declined sharply in number over the last couple of decades, whilst "malicious" vehicle fires continue to increase in number: in the UK at this time they lead to about 20 deaths each year and cost insurers about £77 million per annum. By far the most susceptible type of vehicle is the private car, which accounts for over 85% of vehicles so lost: most of the remainder are light vans, there being few ($\approx$ 3% of the total) heavy commercials. An accidental vehicle fire is most likely to start in the engine compartment where gasoline – well above its flash point at ordinary temperatures – is present. Arson in cars is usually initiated in the interior and an accelerant is commonly used. Vehicles in public service are "fair game" for arson as a form of vandalism. For example, in late 2005 there was an arson attack on a double-decker bus in Brighton, England believed to have been perpetrated by youths in possession of lighters. Emergency responses were effective and none of the 40 occupants of the bus was injured. In Bristol the previous year the upper deck of a bus in service was maliciously ignited.

Mountain mahogany

Botanical name *Cercocarpus montanus*, tree with a remarkable fire resistance attributable in part to its thin foliage. It occurs in California, Oregon, South Dakota, Utah, Texas, Oklahoma and Arizona, as well as in parts of Mexico, including Baja California. It is for many applications a good choice of **firewise plant**.

Mount Barker

This location in the Adelaide Hills, in South Australia[23] was the scene in February 2006 of a fire at an embroidery factory. It is believed that the factory had been broken into and that accelerant had been used in the intentional setting on fire of the building, the damage to which was major.

Moy Park Chicken Factory, Co. Down

The scene in 2006 of a fire originating in an oven. It was attended by 35 fire fighters and seven appliances. The fire was restricted to one part of the factory, destruction of which was therefore a long way from being total.

MSA (Mine Safety Appliances)

This major manufacturer of gas sensing devices, with its HQ in Pittsburgh, includes amongst its products a **photoacoustic infrared sensing** device intended for use in the paint and plastic industries, being directed at substances including toluene and certain ketones and esters. The IR beam is filtered to give the wavelength absorbed by the target compound in each case. MSA make many other types of detector and its photoacoustic device can be backed up by other detectors including one for carbon monoxide in shared cicuitry. (See also **X³ ® Technology, Z Gard® S.**)

Mueller Centurion 250™

Above-ground fire hydrant, having in common with the **American Darling** and the Waterous **Pacer** a 250 psi operating pressure. Like them it has a valve opening of 5.25 inches, although it is also available with a 4.5 inch valve opening. It has three outlets for hose connection, two of 2.5 inch size and one of 4.5 inch size. The Centurion hydrant, with progressive modifications and improvements, has been in manufacture for several decades. Mueller also manufacture **wet barrel fire hydrants**, which are sold with the trade name James Jones.

Mumbai High North

Oil field to the west of India. In July 2005 there was a fire at a production platform which resulted in a death toll in excess of ten. A multi-purpose support (MPS) vessel collided with a riser carrying gas and oil from the well to the platform, which broke open as a result. The leaked contents having ignited, there was escalation eventually leading to platform destruction.

---

23 A very bush fire prone area, the scene in 1983 of part of the "Ash Wednesday" bush fires.

There were 384 persons on the platform at the time. The survivors were evacuated by helicopter and by ship.

Myojo 56 Building

This four-storey building in an "entertainment district" of Tokyo was gutted by fire in September 2001. There were two deaths and a number of injuries. It is believed that the alarm system did not function because it was not connected to electricity and that there were insufficient fire escapes. Many buildings in the same part of Tokyo are known not to comply with fire codes and standards.

200N CiTipel®

Catalytic gas sensor from City Technology in the UK. Described by its manufacturer as a general-purpose pellistor, its performance characteristics are typical of such devices. Assigning its response to methane the value 100 in arbitrary units, its response to propane and to n-butane is 55. Methanol and ethanol cause responses of 85 and 65 respectively. Hydrogen on the same scale has a value of 90 and ammonia a value of 120.

Nagoonberry

Botanical name *Rubus arcticus*, a flowering plant occurring in Alaska and southwards in the "lower 48 states" as far as Wyoming. Native to North America, it is used as a **firewise plant**.

Namdaemun

Historic building in Seoul, South Korea, having been built in the late 14th century. It was extensively damaged by fire in February 2008.

Nanking cherry

Botanical name *Prunus tomentosa*. Shrub native to China and Japan having been naturalised to North America where it is used as a **firewise plant** in places including Michigan.

Napalm

A.k.a. "gellied petroleum", an incendiary substance which as a consolidating thickener uses a blend of salts of *nap*hthenic acid and of p*alm*itic acid. Within this medium there will be hydrocarbons, most commonly gasoline or kerosene. Like other incendiary devices, napalm generates no significant overpressure therefore its action is due to heat release. A further effect, however, is that burning is so rapid that depletion of oxygen for some distance around can be sufficient for there to be an asphyxiation hazard as well as one from carbon monoxide poisoning. Napalm was developed during the Second World War and was used in Vietnam. A Convention of the United Nations from 1980 prohibited use of napalm in civilian locations. Not having been a signatory to the Convention, the US has in fact used napalm in its recent campaigns in Iraq. A "peaceful" application of napalm is the **helitorch**.

Nara

Not only was Nara (then Heijo) Japan's first capital 1300 years ago, but it is also the scene of what might be the longest serving fire hydrant in the world. There is a recent photograph[24] of a fire hydrant at Nara, which when it first entered service doubled up as a post to which to tie horses.

NASFM

National Association of State Fire Marshals. The role and responsibilities of a State Fire Marshall differ between one state of the US and another. State Fire Marshals from all of the 50 states plus the one for the District of Columbia are *ex officio* members of the NASFM. The Association provides services to particular states as requested. For example, it has recently been working with the Maine Fire Chiefs' Association in unifying (making "interoperable") the communications systems of fire departments. Having received Federal aid, this project is now at a pilot stage and is expected to proliferate along the fire departments of the east coast states. The Association also publishes, alone or in conjunction with other bodies such as NIST, reports and commentaries on fire safety matters; such a document recently released by the Association is concerned with the explosion at **Bergenfield, NJ**. It also distributes training material, not all of which originates from the NASFM.

Natik, MA

The scene of delivery in mid -2007 of a new **quint**, built by Pierce to the specifications of the Fire Department, concerning which there have been some initial difficulties. The quint, a top-of-the-range "All Steer" model costing US$657 000, was involved in a crash which the fire fighters' union have attributed to malfunction of the rear steering, although Pierce, having examined the circumstances of the accident, have not themselves endorsed this view. Pierce are nevertheless meeting the cost of repairs. Acquisition of the quint at Natik was in any case controversial in that its ladder and tank facilities are never used, other vehicles fulfilling these roles, yet their presence takes up space which might otherwise be occupied by other rescue equipment. The quint weighs 35 tonnes and its acquisi-

---

24   Go to: http://www.firehydrant.org/pictures/japan.html.

tion necessitated strengthening of a bridge within the area served by the Fire Department.

NATURoil

Vegetable-based hydraulic oil, the flash point of which is specified only as being above 100°C, suggesting safety for most purposes but a lower flash point than comparable products including **Hydro Safe®**. From the same manufacturer (Biochem Wales Ltd) comes a bitumen solvent containing some plant-derived ingredients, for use in cleaning equipment. This too has a flash point of above 100°C.

Negative pressure ventilation (NPV)

Whereas **PPV** has been applied, for example at **Hurst, TX**, to remove smoke so as to prevent it from harming persons and property, NPV is used during "overhaul", that is the process of making the scene of a fire after extinguishment as safe and clean as possible in the circumstances. NPV so applied involves taking air out of the fire damaged building by placing fans at suitable positions within it. Once the fans are positioned fire fighters can make fresh air openings in what remains of the building structure to aid the ventilation.

Neptune LNG terminal

Off Boston, MA, and the proposed scene of usage of a number of offshore support vessels with fire fighting capability at **FiFi 1** level. These will be built by Boston Towing and Transportation (BTT) at a shipyard in Connecticut. BTT entered the LNG tug business after 9/11, when Boston Harbour was seen as a possible target for terrorism. BTT were consequently involved in the building of the LNG tugs *Freedom* and *Liberty*, each of which entered service in 2003.

Neptune P2V

Lockheed aircraft having found application as an **air tanker**. So adapted it has a water holding capacity ≈ 20% less than that of **P3 Orion** or the **C-130 Hercules** (each of which is of course also basically a Lockheed product: the P2V was initially simply the P2). There are a number of Neptune P2V air tankers based at Missoula, MT and also a number with the California Department of Forestry (CDF).

New Brighton, PA

The scene in October 2006 of derailment of a train bearing tank cars holding ethanol. Some of the tank cars finished up in the nearby Beaver

River. Others leaked their contents, and fire and explosion resulted. There were no deaths or injuries, although many buildings had to be evacuated. Ethanol is by now very widely used as a transport fuel in the US, and the accident drew attention to the possible need to move it by pipeline.

Newburgh, NY

Location of a **Good Will Fire Company** having two Pierce pumpers, one of which dates from 1989 and the other from 1998. This is typical of the reliance of such companies on older equipment. The older of the two is used in "mutual aid" situations and is the first of the two out to building fires, whereas the newer one is the first out to car fires.

New Castle, DE

Scene of operation of a **Good Will Fire Company**. It employs career fire fighters and is capitalised largely with Pierce apparatus, some of it a little elderly. For example, the Company operate a 1989 Pierce water tanker which also holds a portable **PPV** device, and two 1991 Pierce appliances of similar specification. It also has a truck with an aerial ladder, manufactured by Pierce and dating from 1996. (*See also* **Newburgh, NY**.)

New London, TX

By the late 1930s natural gas had been distributed by pipeline within the USA for over 50 years, in contrast to the UK where natural gas usage did not begin until the 1960s and used infrastructure originally installed for manufactured fuel gas. Natural gas often occurs with oil ("associated gas") and at the time under discussion this ensured a natural gas supply for Texas, the onshore oil fields of which were producing abundantly (although offshore oil and gas production off the Gulf Coast did not commence until 1945). At a school in New London, TX, a town associated with the oil fields, there was a gas explosion, the death toll from which is said to have been at least 298, mostly of course children. Leaked natural gas had accumulated in one of the rooms and was ignited when an instructor operated an electrical switch which created a spark. A Chevrolet parked close to the school was overturned in the blast. The natural gas which caused the disaster had not been treated with a mercaptan to impart an odour to it. Such treatment became a legal requirement in Texas after the New London school tragedy.

New Orleans, LA, building fire in 2008

Three aerial ladders were amongst the appliances attending this fire at a closed down apartment building in New Orleans during the first few days of 2008. It was classified 4 on the **alarm number** scale. Gas and electricity had been disconnected from the building. The fire fighters managed to prevent spread of the blaze to a nearby church.

Newport, OH

The scene in January 1986 of a fire in a tank of crude oil. After initial extinguishment attempts with water, **frothover** was observed at which stage foam was applied. Extinguishment was total about 40 minutes later.

New Straitsville, OH

The scene of a coal mine fire which was intentionally started in 1884 during a labour dispute involving miners. It has therefore been burning for over 120 years.

New York Fire Department, history of

It was in 1648, one year after Peter Stuyvesant was appointed Director-General of New Netherland, that New Amsterdam had its first Fire Ordinance under which buckets, hoses and ladders were provided to protect the emergent city from fire damage. By 1664, when New Amsterdam became New York, there had been further capitalisation in equipment which included leather buckets to contain water. It was not until 1731 that a fire brigade in the sense of a mobile appliance and crew able to respond to an emergency came into being in the city, with equipment brought from London. Soon after the War of Independence the legislature brought the Fire Department of the City of New York into being. This was operated by volunteers until shortly after the Civil War when the Metropolitan Fire Department came into being with paid fire officers. By this time the steam engine had replaced manual pumping in fire fighting and appliances were horse drawn, having previously been pushed along by able-bodied men.

At this stage the "Metropolitan District" comprised only Manhattan, which was therefore the only part of New York to have the benefit of the Fire Department, volunteer services remaining elsewhere. By 1874 both Brooklyn and Westchester County had become part of the Metropolitan District and the Fire Department had accordingly expanded, although it did not have responsibility for the whole of New York City until 1929, the year the Borough of Queens was added to the Department's coverage area. By then there were motorised fire appliances, radio communications and reticulated water for extinguishment.

Then followed a difficult period for the Fire Department, when expansion was precluded by the depression and later by the Second World War, when materials initially destined for the manufacture of new fire appliances were diverted to war projects. After the war there was urban sprawl in New York.

This and extensive post-war building downtown increased the demands on the Fire Department to the extent that the number of calls it responded to in 1964 was over twice that in 1950. The Department remained well equipped, being one of the first (in 1965) to use the tower ladder. The "Superpumper", capable of discharging 8800 gallons per minute of water, made its world debut in New York that year. Expansion continued and by 1990 the Department was responding to just over a quarter of a million emergency calls a year. A milestone was the first employment of woman fire officers in 1982. In 2007 the Department has over 12 000 uniformed employees and over 2000 civilian ones. It has about 350 mobile appliances of various kinds and responds to over a million calls per year. Nevertheless the Tokyo Fire Department (TFD), which has 18 000 employees and 1800 pieces of mobile plant, exceeds both in number of employees and in capitalisation with plant that of New York. The TFD is in fact the largest such department in the world.

New York State, Office of Fire Prevention and Control

This is part of the New York Department of State, and has a base in the State capital Albany as well as one in New York City. One of its functions is to administer funds allocated by the state legislature to fire fighting purposes. It makes grants available on application to fire departments within the State for the purchase of fire trucks, breathing apparatus and other requisites, priority being given to areas where fire fighting is on a voluntary basis. It also validates in-service training courses for fire fighters in the State and has a program for the prevention of juvenile fire lighting. The State of New York does not have a State Fire Marshal, but a State Fire *Administrator*. The present one is named, appropriately, Floyd A. Madison. Appointed in 2007 on a salary of $105 000, he was previously Fire Chief for Rochester. He is based at the New York City centre of the Office of Fire Prevention and Control. (*See also* **Colorado, Department of Public Safety**.)

NFPA categories of airports

For **ARFF** purposes, airports are categorised on a scale of 1 to 10 according to the largest aircraft which lands there. The NFPA sets requirements for each category in terms of the number of appliances ("ARFF vehicles"), the water release rate and the quantity of foam held at the airport. The Federal Aviation Administration (FAA) has its categories which differ in a few details from those of the NFPA. A Category 9 airport (examples of which include Chicago O'Hare and JFK in New York) must have four suitable ARFF vehicles permanently based at the airport. A Category 6 airport (e.g. that at Bismarck, ND) must have two ARFF vehicles. Categories 1 to

4 need to have one vehicle the size and water release rate of which rises with category number. Airports in Categories 5 and 6 must have two ARFF vehicles and those in Categories 7 and 8 must have three. At the time of its introduction Category 10 was hypothetical, in anticipation of larger aircraft. Now that the double-deck Airbus 380 has entered service, ARFF is having to be closely examined at airports which it uses. Difficulties can arise when an aircraft needing to make an unscheduled landing has to do so at an airport of lower ARFF category than that from which it took off. This happened in 1989 when a DC-10 having taken off from Denver Airport lost its hydraulic functions whilst in the air and diverted to Sioux City Gateway Airport in Iowa. Sioux City Gateway Airport is Category 6, whereas a DC-10 needs Category 8. The DC-10 crashed on its approach to Sioux City Gateway Airport with a resulting fireball and the loss of 111 lives. The disparity in categories of ARFF was raised in the follow-up. (*See also* **Airwolf**.)

Niagara 3-3

Alcohol Resistant Film-Forming FluoroProtein (AR-FFFP) fire fighting foam concentrate. It is used where polar organic liquids such as acetone, ethanol and ether are stored and also at plants where such substances are being made or processed. Its use is not, however, exclusively in fire prevention with oxygenated organics: it is a good foam agent for hydrocarbons generally and is used at offshore installations.

Nigeria, road tanker fire 2007

Notwithstanding its membership of OPEC and the presence of several major international petroleum companies, Nigeria has an appalling record of deaths from oil and gas caused by neglect or sometimes by deliberate sabotage. Theft of fuel is also rife and this of course causes fire hazards. A road tanker carrying gasoline overturned on a badly maintained road in Nigeria in 2007. Ninety-eight persons were killed.

Nimrod aircraft, destruction of by fire in 2006

An RAF Nimrod in service in Afghanistan crashed as a result of fire in 2006, killing 14 persons. It was the worst RAF accident in terms of loss of life since the Falklands War. The Nimrod had just been refuelled in flight from an RAF Tristar, taking on just under 10 tonnes of fuel over a six minute period. Fuel having been received in a tank on one wing was being transferred to a tank on the other. It is possible that the fuel being so transferred came close to pipes bearing hot ($\approx 400\,°C$) exit gas from one of the engines. Another possibility is simply leakage of fuel during the midair refuelling. Difficulties with refuelling Nimrods in the air had been reported previously.

Nippon Dry Chemicals (NDC)

Japanese company manufacturing fire protection products ranging in scale from a small extinguisher to an entire fire truck. Its **carbon dioxide extinguishers** are available in sizes containing from 2 to 3.2 kg of carbon dioxide It also manufactures a system using **liquid carbon dioxide** for installation in ships. It manufactures ABC powder extinguishers in sizes from 3 to 6 kg of the extinguishing agent. Its range of such products is in fact very wide and includes **protein foam** extinguishers, dry chemical extinguishers and devices using **Inergen**. NDC was formed in Japan in 1955 and was merged with Tyco in 2000. The largest such company in Japan, NDC is sometimes commissioned to provide the necessary package of fire protection measures at a major plant, for example at the **Shinagawa Thermal Power Plant**.

Nisshin Maru

Japanese whaling vessel, a "factory vessel" having whale meat processing facilities. In early 2007 whilst it was in Antarctic waters there was a fire on the vessel in which one crew member died. There was concern that fuel leakage from the damaged vessel would follow. The vessel was part of a fleet of six which had been sent to the Antarctic on a whaling expedition. Its propulsion capability was restored whilst it was in Antarctic waters and it returned to Japan for a complete repair. Previously Greenpeace offered the services of a towing vessel so as to get *Nisshin Maru* out of the Antarctic environment as promptly as possible but this was refused by officials in Tokyo. In the event no environmental damage resulted, although the effects on the Japanese whaling industry are expected to be not insignificant.

Nitrile

This is used in the construction of fire hose in three ways. First, a conventional single jacket fire hose can have the further protection of a ribbed layer of nitrile. Suppliers of such hose include the Reliable Fire Equipment Company, which was founded in 1955 and has its HQ in Chicago. Secondly, a single jacket fire hose might consist of smooth nitrile as the inner layer with ribbed nitrile as the outer. Also, as with Reel-Lite 800, a single-jacket hose might have an outer layer of nitrile. A further example amongst many of fire hose made from nitrile is **Guardman**.

Noflan®

Flame retardant for application to fabrics, comprising organophosphorus compounds. Its attachment to fabric is aided by use of a specially developed resin. This enables a piece of fabric having been treated with Noflan® to withstand many washing cycles without loss of the retardant. From the same manufacturer (Firestop Chemicals, UK) comes **Bizon™**. Noflan is finding application to aircraft cabin seats, in competion with **ZirPro®**.

Nomex®

Fire-resistant material used in fireproof clothing, an example of an **aramid fibre** substance. Applications are legion and include fire fighting and motor sport.

Non-maintained emergency lighting

Unlike **maintained emergency lighting** this comes into operation only if there is no mains supply.

North 800 Series™

Range of **SCBA** for fire fighting, features of which include a silicone rubber face piece with polycarbonate for viewing, and a harness made of **aramid fibre** products. It has the facility for "**Buddy breathing**".

Norfolk, VA, nursing home fire

Twelve residents died in this fire in October 1989. The absence of sprinklers was noted in the follow-up and this stimulated moves to have sprinklers installed more widely in the State, e.g. in the **Virginia Beach** hotels.

Northampton Print Finishers

A fire at the premises of this company in May 2007 is believed to have been started by arsonists. There was major damage to plant, including fork lift trucks. Propane stored at the site burnt with considerable overpressures causing blast damage.

North Pyongan

Area of North Korea, the scene of a fire at a gasoline pipeline which led to approximately 110 deaths. The gasoline pipeline was passing though rice paddies where persons were working. However, attempts were being made to appropriate gasoline from a part of the pipeline which was leaking and this undoubtedly contributed to the death toll.

Northumberland Fire Service, acquisition of AEDs by

The fire service for this region in north east England forming a border with Scotland obtained 13 **automatic external defibrillators** in November 2007. Their procurement was from external funding, partly from the British Heart Foundation. In the US and in the UK AEDs for fire services, especially smaller ones, are often obtained in this way. For example the Fire Department at Hall County, GA is currently "soliciting donations" for the purchase of AEDs. One costs about US$3000.

Nozzles, flow patterns from

There are two primary flow patterns available from a fire fighting nozzle receiving water from a hose: the straight stream, a.k.a. the solid stream, and fog, a.k.a. mist. Creation of fog involves the engagement of "teeth" within the nozzle to break up a continuous supply of water into a dispersed one. Hence with some designs of nozzle one flow pattern or another is obtainable according to whether the "teeth" are engaged. Factors such as the pressure of water being equal, the "throw" is not strongly affected by whether such a nozzle is in straight stream or fog mode. The major nozzle manufacturer Akron uses the term **Vari-Nozzle** for a nozzle capable of either spray pattern and field adjustment between the two. A **fog tip** provides for fog release to the exclusion of straight stream. Not only in technical (trade) literature but also in standards and the like "fog" and "spray" appear to be used interchangeably in discussions and descriptions of fire fighting with water. The term "mist" is also appropriate.

NP-30

Catalytic gas sensor manufactured by Nemoto Environmental Technology. When calibrated with methane, it can detect upwards from a concentration equivalent to 1% of the lower flammability limit. It draws 300 mA from a 2 V supply.

NPV

see **negative pressure ventilation**

n-type semiconductor

Often used in gas sensing, such a semiconductor is made by deliberate contamination ("doping") of an element of valency state four with an element of valency state five. The extra electrons provide for electrical conductivity the value of which is affected by contact with gas molecules. From the point of view of gas sensors the most important example is doping of **tin(IV) oxide** with antimony in valency state five.

Nuto H

Series of hydraulic oils produced by ExxonMobil. They are of designation H 10, H 15, H 22 and H 32, where the number is the kinematic viscosity at 40°C in centistokes. The respective flash points are 170, 182, 206 and 212°C.

O'Brien hydrant

One of two fire hydrant types used in New York City and also widely elsewhere in the US. The initial patent on the design is long expired and any manufacturer can bid to supply them. The other fire hydrant used in New York is the Dresser 500 model. Originally developed by the Dallas-based Dresser Industries, this hydrant is now obtainable from other manufacturers including M & H Valve in Alabama.

Ocean Guardian

Vessel for oil drilling in service in the North Sea, the scene in 2007 of a fire which began in one of the engine rooms. Thirty-two persons were evacuated by helicopter, whilst those trained in fire fighting stayed on board and applied their skills with the result that the fire was extinguished after two hours. Importantly, there was no leakage of hydrocarbon from the well which *Ocean Guardian* was drilling. Beside exacerbating the fire hazard this would have led to environmental problems because of entry of crude oil into the sea.

Odin Viking

Anchor handling, tug and supply (AHTS) vessel entering service in 2003 as the *San Fruttuoso* and equipped for fire fighting at **FiFi II** level. The fire fighting plant comprises two pumps each with a capacity of up to 3600 m^3 per hour of water and three **monitors** each with a capacity of up to 2400 m^3 per hour. After having been chartered by an Italian oil company, the vessel served in the North Sea on a "spot market" basis, and is currently in use at the Oudna field off Tunisia.

Ohio Penitentiary fire, 21 April 1930

The worst prison fire in US history, claiming the lives of 322 inmates. The fire began when a candle was brought into contact (probably intentionally, by prisoners planning to escape) with some cleaning rags which had absorbed a flammable liquid. Fatalities were caused both by the heat and by toxic gases. At the time of the fire the Penitentiary housed twice the number of occupants for which it had been built. A positive spin-off to an otherwise appalling event was that a Parole Board was established in Ohio and deserving prisoners selected for early release in order to relieve overcrowding. Thousands of prisoners gained their freedom in this way. The prison camp fire at **Jay, FL** over 35 years later also resulted in some reforms. (See also **Wadi el Natrun Prison**.)

Okinawa, aircraft fire 20 August 2007

A Boeing 737 aircraft operated by the Taiwan airline was destroyed by fire at Okinawa on 20 August 2007: all on board escaped unhurt. A bolt from the wing pierced a fuel line and this was the cause.

Oklahoma, participation of in the Federal Excess Property Program

Oklahoma has been obtaining capital equipment for fire fighting through the **Federal Excess Property Program** since 1959 and since the late 1970s has acquired more equipment through the Program than any other state. As is usual practice, equipment acquired by the State via the USDA Forest Service is passed along to local fire departments. Over that period the number of volunteer fire fighter departments in the State has more than doubled.

Oktyabrskaya station

Part of the Moscow metro system and the scene in 1981 of one of the worst urban railway fires ever. Two railcars were totally engulfed by the fire. The loss of life is not known: all that is known is that western observers counted seven dead persons. The origin of the fire is believed to have been electrical. There was another major fire in the Moscow metro that year, at the Prospekt Mira station.

Oneida, NY

The scene in 2007 of fire and explosion following derailment of a train carrying liquefied petroleum gas (LPG). It is reported that it was also carrying petroleum distillate and toluene.

Ongard®

Fire retardant, containing zinc and magnesium, suitable for materials including PVC, manufactured by Great Lakes. It also finds applications to textiles. It is available in a number of forms denoted by a numeral, Ongard® 2 and so on. Its competitor for use with PVC is antimony trioxide (ATO) over which in *some applications* it has two advantages: it works at lower loads than ATO and acts as a **smoke suppressant** in a way that ATO does not. The emphasis in the previous sentence is intended to remind the reader that a particular PVC product can be affected in its propensity to fire by the nature of the plasticiser. In the presence of some sorts of plasticiser Ongard® and ATO are in fact used together. A "stablemate" of Ongard® is **Smokebloc®**.

Optical methane detector (OMD™)

Natural gas is widely used by electrical utilities, there being enormous trade in natural gas internationally for this and other applications. Both the gas supplier/distributor and the user have responsibilities in early detection of leaks and the optical methane detector is a common choice. It is "classical" in its operation; the instrument supplies infrared light, absorption of which by methane is the basis of detection. Modern forms can be powered by the battery of a vehicle and are therefore very suitable for field use. They are often set up so that the signal from the detector goes to an instrument on the vehicle's dashboard, giving an integrally designed facility.

Oregon grape

Botanical name *Berberis aquilifolium*,[25] this shrub grows along the western side of the US and is a **firewise plant**, owing this characteristic to the fact that it creates little debris.

Oscillating monitor

Monitor such that there is movement through an arc during operation. Energy for the movement is derived from water supplied from the pump to the monitor, a small proportion of which is diverted to drive the monitor movement whilst the remainder is directed at the fire in the usual way. Angular speeds of up to $30°\ s^{-1}$ are possible.

Our Lady of the Angels School

A fire in this school in Chicago in 1958 resulted in the deaths of 92 pupils and also of three nuns who taught at the school. Another 100 persons were seriously injured. The oldest part of the school building dated from 1910 and there had been extensions and modifications, the most recent of which had been in 1951. The building complied with the local fire regulations of the time. These included the use of brick for the exterior walls, although much of the internal structure was made of wood. There were four fire extinguishers

---

25   Some plant taxonomists place the Oregon grape in the genus *Mahonia*.

and a single fire escape. The fire began at 2:20 pm, originating with ignition of the contents of a receptacle for waste cardboard. This released dense black smoke, which drifted to other parts of the building. Escalating effects included breakage of a window by the heat from the smoke, providing air ingress which accelerated combustion. Evacuation began, and students having been directed out of the building by teaching staff were taken into the church for safety. The teachers who re-entered the school discovered that occupants of classrooms on the second floor had become trapped. Some attempted to escape by jumping out of windows and received fatal injuries in doing so. Fire fighters did succeed in rescuing some of those trapped. They were helped by police and also by casual passers-by who used their cars to take injured children to local hospitals. Four years after the fire a boy admitted to having ignited the cardboard waste in which the fire began, but later retracted. It is however clear that whether he had caused the fire or not he was in possession of information on it discovered by the investigators but never released to the public. He was not charged (he is now deceased) and it is not known whether the fire was an act of arson.

Out-of-code building

Building predating building codes for fire protection and not having been upgraded to bring it to within such codes. When an out-of-code building is damaged in a fire and insurance is claimed, it is not in general the responsibility of the insurance company to rebuild so as to bring it within the building codes. This means that it would after repair remain an out-of-code building. Some municipalities require that if the extent of fire damage to an out-of-code building is 50% or more the entire building must be brought within the codes before the building enters re-use. This necessitates extra insurance cover to demolish and rebuild the parts not affected by the fire. (See also **Augusta, ME**.)

OxyReduct®

Fire protection system developed and manufactured by Wagner in which air in a room is recirculated and, by means of a membrane, enriched in nitrogen. An atmosphere of 15% oxygen can in this way be sustained and this inhibits ignition and retards propagation if ignition does occur. There remains the question of how suitable for human occupancy a room with such an atmosphere is. The manufacturers make the interesting point that an atmosphere of 15% oxygen at a total pressure of 1 bar is equivalent in oxygen content to an atmosphere of 21% oxygen at an altitude of 2700 m. The present author has verified this and the calculation is in the box below.

Using the barometric equation:
$$P(h) = P(0)e^{-mgh/RT}$$
$P(0)$ = pressure of the gas at sea level (h = 0), M = molar mass of the gas, g = acceleration due to gravity, h = height above sea level, T = temperature, R = gas constant. Rearranging:
$$h/T = (R/Mg)\ln[P(0)/P(h)]$$
Putting:
$$T \approx 273K, M = 0.0288 \text{ kgmol}^{-1}, g = 9.81 \text{ m s}^{-2} \text{ and}$$
$$P(0)/P(h) = (21/15) = 1.4 \text{ gives:}$$
$$\underline{h = 2700 \text{ m}}$$

The height calculated of 2700 m is much lower than, for example, the altitude of the Colorado ski resort of Winter Park, which is in fact 3676 m. That persons can breathe comfortably under such conditions there is no question and this supports the view that the oxygen reduction brought about by OxyReduct® will not harm persons. Even so, for any particular installation this point has to be addressed and local regulations for minimum oxygen level observed. There is the obvious point that OxyReduct® might be used in storage situations where entry of human beings is only ever brief and infrequent, for example a collection of valuable old manuscripts. (*See also* **British Library book storage facility**.)

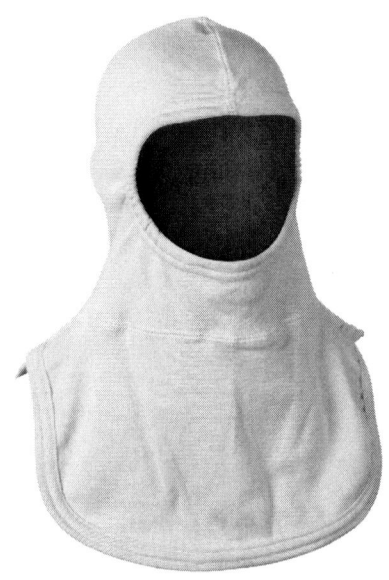

PAC II fire fighter hood (courtesy Majestic Fire Apparel)

P

P3 Orion

This aircraft manufactured by Lockheed includes amongst the many uses to which it has been put aerial fire fighting. An **air tanker** based on the Orion will hold about 3000 gallons of water, which of course can contain an extinguishing foam or a wetting agent. A P3 Orion air tanker crashed over a national park in California in April 2005, killing all three crew members. No actual fire was being fought at the time of the crash: the aircraft was on a training flight. The aircraft, owned by a private company, was to have been on contract with the Forest Service and Interior Department from the following month when the "fire season" began. Many P3 air tankers in use at the present time were previously military or naval aircraft and have been flying for about 50 years.

P140

Conductive carbon substance which has been introduced, typically at levels of 2%, into certain **Nomex®** products (including **Delta™**) as a means of preventing static build-up in fire fighting clothing. Its action has been compared to that of a lightning conductor: it attracts charges from the fabric which are then neutralised by air.

PAC II

Fire fighter hood manufactured by Majestic Fire Apparel in NJ and widely used in the US. It is composed entirely of Nomex®. It has a length, from the top of the wearer's head to the edge of the bib, of 21 inches. The face opening is extensible from 5 inches to 15 inches, according to the muscular movements of the wearer. Being **Nomex®**, it can be laundered or dry cleaned without loss of effectiveness, but not bleached in chlorine. Sister products differ in fit. For example, with some individual fire fighters the 21 inch PAC II leads to "bunching" when worn as part of the total ensemble, and this can be overcome by using the 18 inch PAC IIA.

Paisley, Scotland

The scene on New Year's Eve 1929 of a cinema fire which killed 71 children. In fact, "fire" is possibly a misnomer, as it is believed that there was no ignition. A freshly used spool of film put back into its metal case began to release smoke, which filled the interior of the building and caused the occupants to seek evacuation. The exit door could not be opened, and the victims perished from crushing and from **crowd suffocation**. The process involving the spool was probably largely pyrolysis rather than combustion although, of course, the two are often concurrent and interdependent. It might have been that the children responded to components of the smoke with a very low odour threshold and would not have come to any serious harm had they stayed in their places instead of attempting evacuation with its tragic consequences.

Palembang

Location In Sumatra, Indonesia, the scene in mid-2002 of a fire at a karaoke bar in which 53 persons died. An electrical fault is believed to have started the fire. Fire safety standards in the building were very inadequate.

Panama, bus fire 2006

Eighteen Panamanians lost their lives in a bus fire which began with an overheated electrical cable in the engine. Heat from this was transferred to the air conditioning unit, also in the engine compartment. A safety device which causes the engine of the bus to cut out if the air conditioner overheats would, had it been functional, have prevented serious consequences. This safety device had, however, been illegally overridden. The air conditioner used **HC-12a®** which, being flammable, caused the fire to accelerate. The driver of the bus opened the engine cover, providing enhanced oxygen supply, with the result that the one exit from the bus was blocked by the flames.

"Passive fire protection": a misnomer for intumescent substances?

The terminology whereby a *passive* fire protection device is one which stays in the same place during the fire which it occupied before it, such as a protective door, is almost too well known for inclusion in a volume such as this. By contrast, operation of an *active* fire device does involve movement; for example there is movement of water from a sprinkler. An intumescent substance is classified as passive. The author has, however, noted comments on the Web from experts that without doubt movement is required for a substance displaying **intumescence** to do its job in fire protection, so it is not "passive". This is a valid point, and constitutes an imprecision in the terminology.

"Pacer"

Term often used by fire fighters to mean the **Waterous** WB-67-250 fire hydrant, a descendant of the fire hydrant first produced by Waterous in 1886. The Pacer range has been available since 1996 and has a 5.25 inch valve opening with operating pressure 250 bar. Having noted that **Waterous** and **American Darling** hydrants have been available since the 19th century, we might add that in the US some fire hydrants are still in service over 50 years after installation. The Pacer can be provided with a **Storz coupling** at the outlet.

Paints, flame retardants for

Paint on walls and panelling can form a significant proportion of the fire load of an enclosure, and fire hazards of applied paint applies particularly to corridors and stairwells in apartment blocks. There are ceramic powders which can be blended with a paint to reduce hazards in the event of fire, and "recipes" for paint/retardant blends for a specified degree of retardation according to standard tests. The effect of the powders is physical, involving modification to the properties, including thermal conductivity, as well as increasing the thermal mass of a paint layer.

A noteworthy case of a fire involving applied paint occurred in a stairwell at an apartment block in Bronx, New York in 1996. A few months before the fire a fire-retardant paint had been applied to the walls but the old paint, which had not been fire retarded, had been left underneath instead of being stripped and this burnt with rapid propagation. Fire protective paints might be intumescent, as with **Burn Barrier™ 10-10**.

Pak Tracker™

Fire fighter location device from Scott in the US. In addition to audible and visible indicators it has a radio transmitter the output from which is receivable at a transmitter up to 300 m away. The transmitter can be incorporated into a **SCBA** from the same supplier.

Pan-Am 707, fire in

In November 1973 a Boeing 707 freighter operated by Pan-Am, having departed New York for Glasgow, UK, attempted an unscheduled landing at Logan International Airport in Boston after a fire had occurred on board. By the time the landing was attempted, smoke was preventing the flight crew from operating the controls correctly. It is believed that acid being carried as cargo by the aircraft leaked and reacted with the sawdust packing surrounding it. The aircraft crashed on approach to the runway at Logan and all four crew were killed.

Panther

SCBA from Survivair. Materials used in the face piece of the SCBA include transparent polycarbonate for viewing, silicone rubber for the strap and Kevlar® protective covering. Air supply from the cylinder is monitored continuously and an LED illuminates as warning of a drop in the supply. Fibre glass filled nylon is used in the construction of the backpack. The speaking diaphragm is made from **Kapton®**.

Paraflam

From the manufacturer of **Pyroguard** and of **Pyrostem**, a fire-resistant glass which has a gelatinous inorganic layer between two layers of toughened glass.

Paris, fatal hotel fire in 2005

Accounts of this fire, in which 23 lives were lost, contain one dismal fact after another. The hotel had six storeys yet only one staircase. Many of the occupants were illegal migrants lacking formal documentation establishing their identities. Adding to the intrinsic vulnerability of the building was a wild party held within it at which alcohol and drugs abounded and candles were lit. It appears that contact of the candles with a flammable fabric is what started the fire.

Pasadena, TX, ethylene fire in 1989

Ethylene (C_2H_4) is the most common cracking product and widely used in the petrochemical industry. The Gulf Coast was (along with Germany) the scene of development of cracking technologies about 70 years ago. An accident involving ethylene in one of the Gulf States in 1989 resulted in 22 deaths when at a Phillips Petroleum chemical plant at Pasadena, TX (near Houston) there was an ethylene fire. There were also many non-fatal injuries. Rescue was co-ordinated by Channel Industries Mutual Aid (CIMA, where "Channel" refers to the Houston Ship Channel), which under an established arrangement enables emergency services from a number of districts to respond to such an event.

PASS

In the US, personal alert safety system. Applies both to an **automatic distress signal unit (ADSU)** or to such a device not operating automatically needing initiation by the fire fighter. This might well simply be an "SDU" in the UK. **Pak Tracker™** is one example amongst many others of a PASS. The term "transmitting PASS" is sometimes applied to such.

Patrick Henry Hotel, Roanoke, VA

Declared an **out-of-code building** in 2007 on account of the absence of sprinklers. The owners applied unsuccessfully for an extension to a temporary occupancy certificate until a building permit whereby the hotel would be converted to "senior apartments" was approved.

PBI™

Polybenzimidaxazole fibre, material for use in fire fighter apparel manufactured by Celanese in Dallas and often blended with Kevlar®.

PBY Catalina

PB = Patrol Bomber and Y designates the manufacturer Consolidated Aircraft which, in 1943, merged with Vultree Aircraft to become Convair. The Catalina is a flying boat with floats on the wing undersides, although some were also fitted with retractable landing gear and so were amphibious. Widely used during the Second World War, some were subsequently sold on to be converted to **air tankers** and as recently as the 2003 fire season one such dropped 765 000 gallons of water over the State of Washington during 202 hours of flying time. There is another Catalina **air tanker** in Washington, currently at the airport in Ephrata having recently changed hands.

Pelecanus erythrorhynchos

A.k.a. American white pelican, an adult specimen can weigh up to 10 kg. Such a bird in flight struck a B-1B bomber aircraft over Colorado in 1987. Impact was with the leading edge of one wing of the bomber, and fuel lines and hydraulic fluid lines were severed with resulting fire and destruction of the aircraft. Only three of the six crew managed to eject to safety. (*See also* **JFK Airport, bird strike at in 1975.**)

Pellistor to infrared conversion in gas detection

There has been a tendency to shift from **catalytic gas sensors** (pellistor) to infrared for some applications. A UK company has developed a means whereby the pellistor sensor is removed and replaced with an IR one. An intervening circuit installation makes the potential measuring device previously fitted for the pellistor suitable for the IR detector.

Peloponnese

Area of southern Greece, the scene in August 2007 of forest fires in which the number of fatalities exceeded 40. There was also much destruction of property. At the same time there were fires in other parts of Greece, one of which was close to Athens. The **Mi-26** helicopter was used in extinguishment.

Penhallow Hotel

In Newquay, south west England, the scene of a fire in August 2007 in which there were three fatalities. One of the questions raised as a result of the fire was whether at a resort town such as Newquay there should be increased fire cover over the holiday period. Another was the absence of a sprinkler system from the hotel.

Peoples' Friendship University of Russia

Situated close to Moscow and providing higher education for students from outside Russia since 1961, and the scene of a very serious dormitory fire in 2003 in which at least 40 students were killed. The dormitory was a converted five-storey house.

Peoria heights, IL

Scene of recent delivery of a new **quint** built by **Alexis**, of 100 feet ladder reach. Powered by a 470 hp diesel unit, it has an overall length of 11.8 m and a **Waterous** pump capable of delivering at 1500 gpm and a water/foam tank of 300 gallon volume. It also has power take-off (PTO) generating capability and a hose bed. Alexis have also recently built a quint for use in Milan, IL.[26] In striking yellow livery, this is 13.6 m long and uses a Detroit Diesel engine of 515 hp. The ladder reach is 104 feet. The quint has a Hale pump capable of 2000 gpm and a 300 gallon water tank. It has its own illuminating facilities (such as are often carried by a **rescue vehicle**) of total wattage 1000.

perfFECT™

Product for the disposal of spent fire fighting foam, developed for foams containing **perfluorooctane sulphonate (PFOS)** which has a harmful effect on aquatic life. It works by taking up the foam ingredients on to an adsorbent carbon, which can then be incinerated.

Perfluorooctane sulphonate (PFOS)

An ingredient of fire protection products including **AFFF** as well as other materials including "Scotchguard". It is being phased out in many countries

---

26   Alexis supplies considerable quantities of trucks to its own state and this is true of some other suppliers. It was particularly so with **Pirsch.**

because of its possible harmful effects on humans and wildlife. In related discussions by manufacturers and governments the possible need to make an exception for fire fighting applications was sometimes raised.

Perfluorooctanoic acid (PFOA)

Often an ingredient of **AFFF** as well as being used in paper and cloth treatment. There has been concern that PFOA having entered the environment resists breakdown.

Peshtigo, WI

The scene on 8 October 1871 – the very same day as the **Great Chicago Fire** – of a major fire which claimed over 1000 lives, possibly over 2000 as records were not available to those who attempted to chronicle the tragedy. It is in any case certain that in terms of loss of life it was the worst fire in US history. The blaze began with a prairie fire fanned into acceleration by rapid winds. There were in fact three severe fires that day in the US: the Peshtigo and Chicago ones as noted and also one at Port Huron, MI which killed over 200 persons.

Petrobras platform fire, March 2001

There were 11 fatalities from this fire at an offshore oil production facility off the coast of Brazil operated by Petrobras, and the entire platform was lost. Reports to the company had been made over the previous few days of a difficulty with the pipe work for associated gas and it is thought that such gas had actually found its way into a hollow support member. The accident was the second fatal one involving a Petrobras facility within a matter of weeks. In January that year two persons were killed at a Petrobras production platform for natural gas only. The company is into not only downstream activity but also chemical manufacture and here again safety records are not without serious blemish. In December 1998 three persons were killed in a fire at a Petrobras refinery and in March the same year there were four injuries at a rubber-manufacturing plant owned by the company.

Petroleum products, electrical conductivity of

Though none is a "conductor of electricity" in the conventional sense, petroleum fractions do differ significantly from each other in this regard and this is a factor in handling and storage safety as electrostatic discharge can provide an ignition source. This is believed to have happened in the 2003 accident at **Glenpool, OK**. There, a tank previously having held gasoline, was being filled with diesel. Guidelines apply when such a change is taking place, and these include avoidance of splashing and of excessive flow speeds or other mechanically vigorous operarations which could lead to the creation of charge. What might have happened at Glenpool, OK is that charge was created in the diesel which ignited residual vapours from the gasoline previously contained in the tank. Electrostatic discharge in liquid fuels can be reduced by incorporation of a **fuel conductivity improver**.

Petrol explosion in Shanghai, November 2007

Flying debris from an explosion at a filling station in Shanghai killed a cyclist a considerable distance away. There were three other deaths and persons on a nearby bus were injured. It is reported that maintenance work was taking place at the filling station at the time of the accident.

Philippines, standards of fire safety

There have been many fatal fires in this country attributable to lamentable practices and general neglect. One of the most tragic was at an orphanage in downtown Manila in 1998, in which 25 children and three adults died. Exits from the building, which was made of wood and was 85 years old, were blocked off; windows which might otherwise have enabled some of the occupants to escape had grilles on them. The fire services took a long time to arrive and then brought along only one fire truck. It came to light in the follow-up that the municipal fire department possessed only 40 fire trucks, a mere 15 of which were in anything like serviceable condition. All had been acquired second hand by the department. There are some private fire trucks in Manila owned by businesses and 16 such were sent to the orphanage fire, which was one of a catalogue of fatal fires in Manila. One reason amongst others for the vulnerability of buildings like the orphanage in which the 1998 fire took place is that electrical cables and installations are very old and have lacked periodic checks. A Boeing 737 aircraft belonging to Philippine Air Lines was still on the ground at Manila airport in May 1990 when there was an explosion in one of the fuel tanks and a knock-on fire involving the aircraft cabin. Electrical heating through damaged wiring is believed to have been the cause. Eight passengers were killed. On the positive side, we note that a very advanced **ARFF** vehicle – the Ziegler **Z8** – is in service at the airport at Mactan (Cebu). (See also **Quezon City, hotel fire**.)

Phoenix

Fire boat belonging to the San Francisco Fire Department, entering service in 1955. One of her most notable performances was during the 1989 earthquakes. In the part of San Francisco known as the Marina there was a post-earthquake fire to which the *Phoenix* was summoned. From close to shore, 6400 gpm of water were directed from the *Phoenix* to the areas affected by fire over 15 hours. The vessel was carrying sufficient fuel to power its pumps, which were supplemented by two pumps at an engine on shore, for this protracted period. Replenishment of pump engine fuel by means of a tug and supply vessel would otherwise have taken place as a routine operation.

Phos-Chek® AquaGel

Marketed in Europe as Phos-Chek® Focstop, a gel fire fighting agent performing similarly to **Thermo-Gel®** and **AFG Firewall™**, although unlike either of those it is supplied as a powder. BASF had a part in its development. Like other such gels it is used in extinguishment in forest fires.

Photoacoustic infrared sensing

The most advanced application of infrared to gas sensing. Infrared at a wavelength known to be absorbed by the target gas is applied. If the target gas is present, absorption takes place, the temperature and pressure of the bulk gas containing the target gas are both raised, and the pressure rise leads to an acoustic effect. This is detected and amplified. Photoacoustic infrared sensing has found application in the petrochemical industry.

Photodiode, use of in smoke detectors

A photodiode is a device which on receiving electromagnetic radiation responds with an electrical effect. When it is used in smoke detectors, the set-up is such that infrared light passing in front of the photodiode is, in the presence of smoke, scattered so as to impinge upon the photodiode and this is the basis of the signal. Alternatively, the photodiode might be receiving infrared light continually and respond to a *reduction* in its intensity when smoke enters the beam and diverts some by scattering. A smoke detector of this sort is often referred to as a photoelectric smoke detector or as an optical smoke detector.

Photoelectric cell

In fire engineering, the importance of a photoelectric cell is that it is the basis of a UV flame detector. The cathode in such a photoelectric cell releases electrons when UV light is incident upon it, and movement of these to the anode creates a detector current. An IR flame detector works similarly. Sometimes a flame detector will have one cell for UV and one for IR.

Pierce Quantum

Fire apparatus from Pierce Manufacturing whose HQ is in Appleton, WI. It is often configured as a **quint**. Its aerial ladder reach is 105 feet and it carries 1200 feet of hose. Pierce manufacture a wide range of fire fighting appliances with aerial ladders, some of which have a platform (**ladder bucket**). Generally Pierce have a policy of using steel for aerial ladder construction. However, the one of lowest specification has 75 feet reach with telescoping *aluminium* ladders. One such is in service with the Honolulu Fire Department. Other apparatus from Pierce Manufacturing includes the **Contender**. The Las Vegas Fire Department have replaced their entire fleet of 28 fire vehicles with appliances based on the Quantum, only some of which of course are "aerials".

Piezoelectric generation of sound

For sound to be so generated involves no moving parts and therefore very low electrical power. Many fire alarm devices work by this means by drawing current at milliamp level. The contrasting means is eletromechanical generation, as with the classic **Q-siren**, which requires electrical power in excess of a kilowatt.

Pinewood, NC

Scene of recent delivery of a **rescue vehicle** based on the **Ford F-550**, though having a 325 hp diesel engine made by International. The vehicle has flood lights of combined illumination 1950 W. The Long Creek Volunteer Fire Department, also in North Carolina, recently received an ambulance also based on the Ford F-550 from the same vehicle builder in Wisconsin that supplied the rescue vehicle to Pinewood.

"Ping pong ball"

Term used by fire fighters to mean a plastic sphere treated with a suitable organic liquid (often glycerol) and potassium permanganate solution. These are dropped from a helicopter and their subsequent ignition provides a means of **controlled burning** of forest or grassland. They therefore fulfil the same function as a **helitorch** over which they have the following safety advantage. With the helitorch ignition of the napalm occurs *in situ*, whereas with the "ping pong" ball ignition is delayed and when it occurs cannot in any way endanger the helicopter or its crew.

Pinzgauer

Truck, designed and initially manufactured by Steyr-Daimler-Puch in Austria and later made under licence in the UK by British Aerospace, often adapted to fire truck application. For example, eight Pinzgauer-based fire trucks are in service in the Isle of Man, each with a water tank, pump and 9 metre ladders. In service in England is a "light pumper" made by **Angloco** using a Pinzgauer vehicular base. This sort of "hybridisation" is very common in fire appliance building.

Pirsch

Manufacturer of fire appliances from the 1920s to the 1980s, based in Kenosha, WI. A few have survived into the 21st century, including a restored **tractor-drawn aerial** in Dallas, TX. The last Pirsch appliance to be "retired" in Kenosha itself was a pumper, manufactured in 1984 and in use until 2007, although for the last six years its use was restricted to training.

Piston engined aircraft, fuel for

These use fuel in the gasoline boiling range. Their extremely low flash point (well below room temperature) makes for hazards on collision. In June 2006 a Beech Baron aircraft crashed having taken off from Corpus Christi, TX for a destination in Louisiana. Someone inside a mobile home at the scene of the crash was killed by the resulting fire.

Places of worship, arson attacks on

In the UK every year there are hundreds of arson attacks on such buildings. One obvious reason is that at churches and the like funds for maintenance and protection are seldom abundant so security measures are not state-of-the-art. Thieves breaking into church buildings for articles of value such as sound systems, easily converted to cash, might attempt to destroy evidence by torching the building once they have removed what they want. There are those with a compulsion to damage and destroy property and for them a church or church hall might be an easy target. The Salvation Army have lost buildings in several parts of the world through arson. For example, a Salvation Army store in South Lebanon, OH, which contained food and used clothing for the needy, was destroyed by arson in 1998. At least some of the refuges for the homeless run by the Salvation Army will not admit someone known to have a conviction for arson.

At Lincoln Cathedral, England in 2002 there was an intentionally created fire in which prayer candles had been the tool of the arsonist, these having been lit and placed in contact with flammable fabric. Damage was about £0.25 million. There is concern widely amongst cathedral administrators that plastic seating, which is a fairly recent introduction to such buildings, will make them more vulnerable in the event of fire initiation. Westminster Cathedral in London has recently had **XP95** detectors fitted.

Platte, SD

This has a rating of 6 on the **Public Protection Classification (PPC™)** scale. There are no class 1 communities in South Dakota and only a few with a rating as good as or better than 6. To say that "6 is good for South Dakota" would really be a misrepresentation, for two reasons. First, the Insurance

Services Office is not in any sense an official regulator of fire services: it merely advises member insurance companies. Secondly, the insurance benefits accruing from the PPC™ rating vary from state to state and depend on factors including the population distribution and the terrain. It might be that the costs involved in extending and improving a fire department in an attempt to rise say from class 6 to class 5 would colossally exceed insurance benefits and would not be a responsible use of financial and other resources. (See also **Selected US states, Insurance Services Office classifications for**.)

Plenum cables, fire hazards due to

Plenum cables are cables installed behind walls, under floors and above ceilings and are much more numerous now than before the "IT revolution". They have the nature of communication cables, enabling a voltage which signifies a quantity to be measured. The current within such a cable will normally be minuscule and the danger is not from electrical heating. The danger is simply that the insulating material on the plenum cables adds to the fire load of a room. Teflon is therefore a common choice of insulator for such cables, being fire resistant: PVC is also used. (See also **Fire pillow**.)

Pneumatic sensor

Alternative term for a device using **photoacoustic infrared sensing**. Also an alternative term for a device using a **metal hydride actuator**. When the term is encountered therefore, it is important to clarify in which sense it is being used. Both devices work by creation of gas pressure, in the former case by infrared absorption and in the latter by a chemical decomposition. It is therefore unsurprising that this ambiguity has slipped into the "language" of fire engineering. Often, an aircraft engine is said to contain a pneumatic sensor, and in that context the term always means a metal hydride device.

Podolsk

This location near Moscow was the scene in late 2007 of a fire at an adhesives factory which injured seven fire fighters. Whilst they were tackling the blaze to which they had been summoned, a nearby container of combustible material burst open and the contents ignited.

Pokador 500/750

Family of **fixed gallonage nozzles** manufactured by LineGear, Santa Margarita, CA with water release capability, depending on the model, between 40 and 300 gpm. Each model in the range can be used in "straight stream" or "fog mode". Terminology in the trade literature appears to be that it is axiomatic that a fixed gallonage nozzle has this capability. If it is set up for fixed gallonage with straight stream only, it is more often called a "tip only". The Pokodar 150 LP, also a LineGear product, is an example of a "tip only".

Poland, recent electrical fires

Three house fires in a single weekend – in Pruszkow, in Lodz and in Kalisz – led to the loss of eight lives in Poland recently. Overloading of electrical circuits, partly with heating devices drawing heavy current, had caused the fires. Such overloading had led to cable heating and insulation failure.

Polycarbonate, construction of nozzles from

This very robust polymer material can be used to fabricate nozzles for fire fighting; as described in other entries, brass or aluminium are the alternatives. Some nozzles made from polycarbonate are for straight stream release only, whilst some are capable of fog formation. Polycarbonate nozzles are available which are built to withstand the pressures required by a **booster fire hose**. Nozzles made from polycarbonate have the advantage of very low cost: one capable of straight stream and fog release can be purchased for less than US$S20 (2007 rate).

Polyester/cotton blends, dangers of

When a polyester/cotton blend is ignited, there is a tendency for the cotton alone to burn and for the polyester to provide a support structure for the burning material. This is noted in nightwear regulations in the UK. Some concern has been expressed that upholstered baskets for holding a baby ("Moses baskets"[27]) made of such a blend are on the market.

Polyetheretherketone (PEEK)

Polymer substance with good fire resistance. Its many uses include aircraft noise and thermal insulation. An example of a PEEK product having found such use is Lamaguard™130MD covering film, manufactured by the Lamart Corporation in Clifton, NJ. It is supplied as a roll. Boeing use Lamaguard™130MD in some of their aircraft.

Polyurethane foams, mine accidents involving

Polyurethane has found application to the mining of precious metals and there were two fires at such mines, one in the 1970s and one in the 1980s,

---

27 The original "Moses basket" – that which held Moses himself as described in Exodus – is thought to have been smeared with crude oil to make it water repellent!

where the burning of polyurethane is believed to have contributed to the fatal effects. At Sunshine Mine, ID, no longer being worked but once one of the major silver mines of the US, there was a fire in 1972 in which 99 lives were lost. A vertical support structure within the mine, constructed of timber, had been covered with polyurethane foam believed at the time of installation to be fire resistant. A fire, the origin of which was never unequivocally determined, began in the mine and when it reached the polyurethane layer, that burnt with abundant smoke release. It is believed that the polyurethane foam exacerbated the mine fire in two ways: being of low thermal resistance it burnt rapidly and accelerated the fire, readily igniting timber with which it was in contact, *and* it led to smoke and carbon monoxide production.

A commentator on the Sunshine Mine fire has made the point that a mine fire 14 years later also involving polyurethane foam might not have happened if the lesson from Sunshine Mine had been properly acted upon in the industry. At the Kinross Gold Mine in South Africa in September 1986, 177 lives were lost in a fire that began with a welding operation. Parts of the internal walls of the mine were covered with polyurethane and this readily ignited and burnt rapidly with abundant smoke.

Polyvinyl chloride (PVC), use of as an insulator for electrical cables

PVC burns according to:

$$[CH_2CHCl]_n + (3n/2)O_2 \rightarrow 2nCO_2 + nHCl + nH_2O$$

It is resistant to ignition and to combustion propagation after ignition. PVC into which no additive or filler has been incorporated requires a temperature well in excess of 300°C for ignition. If ignition occurs, propagation is precluded by the very high oxygen requirement. Whereas most polymer materials will burn in atmospheres down to about 17% oxygen, an oxygen concentration of about 50% is required for sustained PVC combustion. This is of course over twice the oxygen content of air. These properties make PVC an attractive choice for electrical cables; in the event of electrical heating due to overloading, a cable insulated with PVC will display a high degree of fire resistance.

PVC as produced is a rigid material and needs treating with a plasticiser for use as insulating cover of a flexible cable. There are several such plasticisers, many of them derivatives of phthalic acid or of adipic acid. These can in some degree impair the fire resistance as on heating they will release flammable decomposition products. However, it is possible for this to be counteracted by chlorine atoms released by the primary constituent PVC, which will scavenge reactive intermediates so that in effect the PVC itself is a flame retardant. (*See also* **uPVC**.)

Porsche factory explosion, 2008

There were two injuries when at the Porsche factory near Stuttgart there was an explosion which began in the area of the factory where paint drying takes place. In addition to the damage from the explosion, there was significant non-thermal damage from the extinguishment water.

Portable fire pump

In addition to a **vehicle-mounted pump**, a crew attending a fire will often use a portable fire pump. All modern portable pumps for fire fighting purposes are of centrifugal design. The pump will be carried in a locker on a piece of fire apparatus. The pump will be powered by its own built-in gasoline or diesel engine and will be carried manually from the appliance to where it is required on site. Because of the need to keep the weight down, engine blocks are sometimes made of aluminium. One example of the portable fire pump is the **Powerflow 8/5** (UK manufacture): others are the **Firefighter® 5 Series** (Australian manufacture) and the **KTH-50X** (Japanese manufacture).

Port Arthur, TX, refinery fire

Electrical power failure is believed to have led to a fire in the hydrotreating unit at Valero Energy's Port Arthur refinery recently. Stoppage of liquid flow when the power failed led to build-up and loss of containment, which was followed by ignition of the vapours.

Porte d'Italie

Location on the Paris metro system where, in March 1973, there was a fire caused by malicious ignition of a seat in a train carriage. There were two deaths from smoke inhalation. There were two other major fires in the Paris metro system in that decade, neither involving loss of life. In March 1977 there was a fire originating in electrical cables at Charles de Gaulle-Etoile station and in March 1979 another electrical fire at Reuilly-Diderot station. There was also a fire in the Paris metro system in 1985, when at the Barbès-Rochechouart station rubbish which had ignited set fire to cables. There was copious release of toxic smoke and six persons were treated for inhalation.

Port Khalid, UAE

One of the major ports of the United Arab Emirates, and the scene in August 2007 of a fire which resulted in the hospitalisation of three fire fighters having experienced smoke inhalation. The fire is believed to have begun in a small factory and then involved a lubrication oil storage facility, where acceleration of the fire occurred. There were major disruptions to operations at the Port.

Port Wentworth, GA

Scene of a recent (early 2008) fire at a sugar refinery in which there were six deaths and several serious burn injuries. Sugar is one of those materials, often seen as innocuous, which can in fact ignite as a dust or powder suspended in air. (The same is true of certain pharmaceuticals.) It is believed that that the Port Wentworth fire began that way, escalation following. The behaviour of sugar in a fire can be understood from similarity to the elementary chemistry experiment in which sulphuric acid is added to sucrose with the result that it dehydrates and forms a solid mass. In a fire involving sugar this solid mass is highly dangerous, being capable of inflicting severe burns on persons and resistant to cooling by water application. Comparisons which have been made with volcanic lava are not totally without validity. Water was dropped from the air on to the Port Wentworth refinery fire.

Positive pressure ventilation (PPV)

In fire fighting this term means the use of fans, situated outside the building where the fire is taking place, in order to aid the removal of smoke from the building. As well as having a cooling effect through loss of heat in the smoke, this also reduces the likelihood that there will be victims of smoke toxicity. A bonus is that visibility is improved, to the advantage of the fire fighters. There have been reservations expressed by organisations including NIST. The obvious difficulty is that air ingress brought about by PPV could accelerate a fire; it is possible in a particular fire that this and the cooling effect will be very finely balanced. There are certain types of fire, however, for which the advantages of PPV are fairly clear, notably the tunnel fire. Airport fire crews often have provision for PPV. Those who might deprecate use of PPV *during* a fire agree on its value when a fire has been extinguished but the smoke from it has not escaped and remains a threat. There was an example of this in **Hurst, TX**, where a fire in one shop in a mall had been put out but the smoke had drifted into other parts of the mall. A device used by a fire department to provide for PPV is called a pressure ventilation unit. The vehicle transporting it is a mobile ventilation unit (MVU). There are some in the fire protection profession who take the view that more widespread use of PPV in fire fighting awaits developments in ventilator design.

Potassium acetate

Classical fire extinguishing agent. Although its use has declined, there has been a revival of potassium acetate extinguishers for use in restaurants in particular. Such an extinguishing agent, in which the potassium acetate solution has been adjusted to a low pH, is manufactured by the Amerex Corporation in Alabama, who supply it in 6-litre quantities in hand-held extinguishers.

Potassium bicarbonate

Dry extinguishing agent the primary action of which is the release of carbon dioxide, there also being a significant contribution from radical neutralisation as with such extinguishing agents as **BC30**. Other things being equal, potassium bicarbonate significantly outperforms sodium bicarbonate in fire extinguishment. This discovery led to the development and commercialisation of **Purple K**.

Powerflow 8/5

Portable fire pump from the Godiva stable. Depending on the pressure, it delivers water at rates in the approximate range 200 to 300 gpm. It weighs 70 kg and can operate on a single tank of fuel (gasoline) for 90 minutes.

PowerFlow boom turret

Manufactured by Colet in California, a **turret** mounted on a mast ("boom") atop an **ARFF** vehicle. The mast can be moved up and down, forwards and backwards, and horizontally through 360°, giving a very wide range of possible orientation for precise direction of extinguishant. It is available on the **Colet Jaguar** as well as on other Colet ARFF ranges.

PPV

see **positive pressure ventilation**

PPV fans, hazard with

It happened in Canada in *circa* 1994 that in the use of a gasoline-powered **PPV** fan at a fire the fan blades, made of aluminium, broke and fragments escaped through the safety screen and lodged in the ceiling. The Canadian Ministry of Labour accordingly issued an Alert. This identified the make and model of the fan and quoted from the Occupational Health and Safety Act, there being no legislation specific to such devices. This incident is invoked by the manufacturers of fans of the **Ventry®** type, which have non-metallic blades which will disintegrate into harmless small pieces in the event of fan blade failure.

PR1005L

Sulphur compounds in jet fuel are a factor in corrosion necessitating protection of the inside surface. PR1005L is a lining material for integrally designed fuel tanks on aircraft. Applied as a viscous liquid, it forms a hard film after solvent evaporation.

Premier Foods, Bury St. Edmunds

The scene in October 2004 of a major fire. Production was resumed as quickly as possible in order to meet Christmas production targets, and this involved use of a mobile temporary water softening unit until the one which had been destroyed in the fire could be replaced. The factory requires 30 tonnes per hour of softened water.

Pressurised bladder

Simple and widely used type of proportioner for the preparation of foams in fire fighting. Water for extinguishment is passed through a valve which diverts a small fraction of the flow past a "bladder", that is a flexible container, containing foam concentrate. This creates a pressure at its surface, which causes concentrate to exit the bladder where it is mixed with water downstream of the valve.

Prisoners, deployment of in fire fighting in California

In California there is provision for prison inmates of either sex to help in fire fighting and in so doing earn some reduction in sentence. Prisoners so selected are given professional training in fire fighting and organised into groups. In the severe forest fires in California in the fall of 2007, 3000 prisoners took part in fighting the fires along with 6000 "free" participants. Amongst the tasks they were assigned was clearing plants in order to create an area of low fire load. They might also collect water. They are distinguished from regular fire fighters by wearing an orange outfit. The 1999 and 2000 fire seasons in California each saw the death of one prisoner fire fighter.

Proban®

Term applied to fabric having been subjected to a particular chemical process to make it fireproof. The process was patented by Rhodia in 1955, and is sometimes carried out under licence by other companies. It involves introduction of a monomer into the fabric followed by ammonia treatment. This brings about cross-linkage, forming a polymer structure within the fabric. The final step is neutralisation which consolidates the polymerisation and removes by-products. The process was initially developed for cotton, but cotton/polyester blends and synthetics have since come within its scope. Treated fabrics on exposure to heat form a char layer. The fabric does not smoulder or melt.

First applied to children's sleepwear, Proban® has since been used for protective wear in many industries. It is also used in upholstery, mattresses and soft furnishings and in protective clothing for the motor sports. "Proban® finished fabrics" retain their fireproof properties on cleaning and laundering subject to certain conditions and precautions. Dry cleaning is not a problem: it does not attack the polymer structure built up in the processing. Washing in water requires suitable choice of detergent and also reasonably soft water. Chlorine bleaches must not be used as these *do* attack the polymer: if bleaching is required, hydrogen peroxide should be used. Proban® products are diverse and available worldwide, consequently there is a need jealously to protect the label. At the time of writing non-genuine products with a label difficult to distinguish from the Proban® one are coming out of China, and Rhodia have initiated legal action.

Pro-Lite

Fire hose manufactured by Key Fire Hose Corporation,[28] constructed of woven synthetic fibre encased by **nitrile** and pressure tested up to 600 p.s.i. Pro-Lite comes in diameters of 1.75 and 2.5 inches. The sister product LDH-600, made of the same materials in the same configuration, comes in diameters of 4 inches and 5 inches. Between the 1.75 inch and 5 inch forms there is a difference of a factor of 6.25 in the weight per unit length. That is why Pro-Lite rather than LDH-600 is recommended for high-rise use.

Promask PP

We intuitively expect the face mask to be the part of a **SCBA** with the most scope for safety-enhancing development. The Promask PP from Wormald has a "drinking port", enabling a fire fighter to take in liquid without removing the mask. This of course is very helpful in overcoming the effects of fire fighter dehydration.

Promastop UniCollar

Fire collar for use with PVC, developed and manufactured by Promat Fyreguard in Australia. Intended for retrofitting, it is installed by wrapping strips of the intumescent material around the pipe and securing it with brackets. (*See also* **uPVC**.)

---

28   Based in Miami, FL, and the world's largest manufacturer of fire hose.

Propane deasphalting, fire during in 2007

Components of crude oil of higher boiling range than diesel can be improved by deasphalting with a solvent, and propane has found application as such a solvent. At the McKee Refinery in Sunray, TX in 2007 there was an explosion at the propane deasphalting unit which caused serious injuries to three persons.

Propane detector

A common form of propane dectector uses a tin dioxide sensor like some carbon monoxide detectors do. In **mobile homes** and in recreational vehicles such as caravans and campervans, LPG will be used for cooking. The danger from leakage is such that in very many parts of the world installation of an approved propane detector in a campervan is mandatory, and it is required to activate when the propane concentration is a quarter of the LFL (lower flammable limit). Note that there are detectors which respond to propane amongst other gases including carbon monoxide: the **TGS 813** is an example.

Prospect Heights, IL

The scene in December 2006 of a building fire classified by the fire department which responded as 4 on the **alarm number** scale. Eight vehicles attended the fire, including two pumpers. Some deficiencies in the building from the fire safety angle were noted in the follow-up, in particular the fact that the firewalls did not extend far enough.

Protein foam

A widely used extinguishing medium for some kinds of fire, one application being to crude oil fires or to fires involving heavy fuel oil composed of refining residue. The protein constituent is fibre protein resembling the keratin in wool, and is often obtained as poultry feathers. Such a foam works by covering a fire and excluding oxygen. There are many proprietary "foam concentrates", which are combined with water at concentrations of around 5% in order to prepare a protein foam for use at a fire. Whilst suitable for hydrocarbon fires as described, a protein foam might be less suitable for a fire involving an oxygenated hydrocarbon such as an alcohol as such materials can break down the foam. There are fluorinated substances which are alcohol repellent and to include one such in the concentrate can overcome this difficulty, in which case the foam is termed a fluoroprotein foam (FP). An example is **Fluoromac-Plus AR**.

PT-1250W

One of a range of piezoelectric **sounders** from Malory Sonalert in the US. It releases noise at about 5 kHz and draws 1.5 mA when receiving an AC voltage of 5 V peak-to-peak. The sound is at 80 dB level. The sounder is external to the smoke or heat alarm to which it responds. This is fairly common although some, including the **FSA-410BS**, have a piezoelectric system internally fitted.

Public Protection Classification (PPC™)

An activity of the Insurance Services Office[29] in the US. The Office collects information from fire departments and on the basis of such information assigns a particular area a classification on a scale from 1 to 10, 1 denoting outstandingly good fire protection and 10 denoting a level of protection below the minimum set by the Insurance Services Office. These classifications are noted by the insurance companies in setting premiums, so a good rating makes for an advantageous premium. The classification assigned to a particular place ("community") depends upon factors including the staffing and equipment of the fire department and the number and condition of the hydrants. Records of response times are also taken into consideration. A classification, once awarded, is reviewed periodically and can be adjusted up or down. In the latter case the fire department is given 90 days to implement an improvement program before the lower rating is ratified. The Public Protection Classification is used in some degree in all states of the US (not only the "lower 48") although New York City and DC are unclassified. To be classified as 1 is challenging and many states do not have a single class 1 community within their boundaries. This is true of all the states in the quadrilateral obtained by joining on a map Washington and Oregon in the west and the Dakotas in the east, also of states including Arizona, New Mexico, Kansas and Colorado. Class 1 communities are distributed from the north east (e.g. Hartford, CT, Cambridge, MA) south to Florida and as far west as Texas (e.g. **Mesquite**) also in California, particularly in the south (e.g. Anaheim).

Pulaski axe

Tool used in the fighting of vegetation fires, sometimes to create a fire break whereby some trees are cleared to interrupt combustion propagation. A Pulaski axe (a.k.a. a paluski, note the lower case) has two heads on the end of a single shaft at the same level along it. One is a simple axe and the other a mattock, that is a pick axe having several prongs.

---

29 Often abbreviated to ISO, in which case it should not be mistaken for the standards body of that name.

Invented in 1911 by a ranger with the US Forest Service, the Pulaski axe is available today and suppliers include Chubb. Fire fighters at **Mann Gulch, MT** were initially in possession of such devices but later on orders jettisoned them.

"Pumping-Nurse"

Operationally similar to "**Gravity Nurse**" in that the **attack vehicle** is supplied with water from a **water tender**, with the difference that such a supply is not by gravity but by a pump. It is more common in rural fire fighting than the "Gravity-Nurse" approach. Provided that **dump tanks** are being carried, a change from pumping-nurse to **dump-and run** can be made if water requirements are higher than initially expected.

Pumps for sprinklers, horse power of

Amongst sprinkler manufacturers and installers difficulty in correlating pump horse power with protected area in any reliable predictive way is acknowledged. A first-stage estimate is however obtainable from:

$$\text{Pump power (hp)} = \text{Protected area (square feet)} \times 0.0012$$

On this basis a 100 hp pump would protect an area of about 80 000 ft^2. It is emphasised that the above equation is no more than the roughest guide and that a user of it should not expect it to stand up in law in the event of litigation relating to a particular sprinkler system.

Purple K

Dry extinguishing material containing 92% of **potassium bicarbonate**, in widespread use for over 40 years. Typically, particle sizes are <355 μm, some being as small as 60 μm. Its very many uses include aircraft accident fire fighting, for which it is the only dry chemical agent approved by NFPA. Purple K 80 is a similar product, containing less potassium bicarbonate (80% instead of 92%).

Pyranova®

Fire-resistant glass owing such resistance to its structure, whereby a layer of clear intumescent material is placed between two panes of glass to form a laminated product. The intumescent material will provide for integrity even if the glass at one or both sides of it breaks. Pyranova® is a product of Schott whose HQ are in Mainz, Germany.

Pyrodur®

Fire-resistant glass from Pilkington. It has a single layer of intumescent material between the layers of glass. It has recently been applied to the building of Terminal B at Dusseldorf Airport, along with its sister product Pyrostop®.

Pyroguard

Range of fire-resistant glasses developed and manufactured by CGI in Merseyside, UK. It can be supplied as **wired glass** or, without impairment of fire or impact resistance, in wire-free form. It owes its properties to a resin substance which is included in the laminated structure of the glass. Wire-free Pyroguard is available in 7.2 mm and 11.4 mm thicknesses, with performances under simulated fire conditions of respectively 30 minutes and 60 minutes before failure. Fire doors in *all* of the schools in the Isle of Man are to be retrofitted with the 60-minute Pyroguard glass. CGI's other products include **Paraflam** and **Pyrostem**. (*See also* **Heart and Soul Restaurant**.)

Pyrophyte

Term to some extent synonymous with **firewise plant**, denoting a tree or shrub with a high degree of resistance to fire. Such trees tend to have a high moisture content and/or thick bark. The well known California redwood is a pyrophyte, which is one reason why some specimens of it are over a thousand years old, having resisted destruction by fire over that time span.

Pyrophoric scale

The often very significant sulphur content of crude oil can over time lead to the formation of iron sulphide on pipe and tank surfaces. Iron oxide reacts exothermically with moist air and heat so produced can provide an ignition source for vapours. Iron oxide having so deposited is therefore termed pyrophoric scale, and can be removed by washing or chemically with potassium permanganate. There was a serious explosion at the refinery in Grangemouth, Scotland in 1987. In the report on it pyrophoric scale was mentioned as a possible contributory factor.

PyroRope™

Material composed principally of **E-glass**, used for the fabrication of components including gaskets which have to withstand moderately high temperatures – up to 260°C – for long periods. A coating of silicone rubber affords surface smoothness and prevents fraying of the fibres. This coating also makes PyroRope™ resistant to chemical attack by aqueous or organic agents.

Pyrosealant™

Heat-resistant adhesive for use at temperatures up to about 285°C. Composed predominantly of silica, it owes its red colour to its iron oxide content. It was developed and is manufactured by ADL Insulflex Inc., whose other lines include **PyroRope™** and **Silicaflex™**.

Pyrosleeve

Fire collar range from Pyropanel in Australia. It includes the collars which have to be installed at the same time as the pipe which it protects as well as collars which (like the **Promastop UniCollar**) can be retrofitted to a pipe already installed. Across the range fire ratings, measured according to widely used standards, from 2 to 4 hours are available.

Pyrostem

From the manufacturer of **Pyroguard**, **wired glass** with a good degree of fire resistance and of impact resistance. The material was developed for buildings already having wired glass doors of an earlier generation where the doors needed replacing but it was desired to conserve the "ambience". It is polished as supplied.

Pyrovatex CP

Substance for use in making cellulosic substances fire resistant, manufactured by Ciba-Geigy. In addition to the organophosphorus reagent it contains **melamine**. In application to cotton the melamine and the organophosphorus compound react together by condensation, that is combination of molecular structures with the elimination of a simple molecule, in this case water. There is further condensation at the final heating stage when the "simple molecule" by-product is formaldehyde, which is therefore present in the effluent from the process as is methanol similarly formed.

Q

Q-siren

Manufactured by Federal Signal and very common on fire trucks in the US. It works eletromechanically and delivers up to 123 dB.

Quaking aspen

Botanical name *Populus tremuloides*, tree species suitable for fire protection and used in that way in states of the US including Colorado and Montana. The shedding of needles, leaves and dead branches ("debris") is a factor making a particular tree unsuitable for such use as this provides ground-level fuel for a fire. Quaking aspen is noted for shedding little debris.

Quad

Approximate synonym for **quint**, though a much less wisely used term. Logically one would expect a quad to have one fewer functions than a **quint** but it is doubtful whether in common word usage this is so.

Queen Elizabeth 1 (QE1), destruction of by fire

Following her replacement by QE2 in 1969, QE1 was taken first of all to the Florida coast for conversion to a hotel and, after that had been a failure, to Hong Kong harbour. There she was extensively damaged by fire, believed to have been started intentionally, in early 1972. The wreck remains at the bottom of Hong Kong harbour.

Queen Mary 2 (QM2), fire protection systems on

This vessel, launched in 2004, can carry up to 3000 passengers and is used on transatlantic routes. Fire protection is by a **Hi-Fog®** system, which enables water dispersed into fine droplets to be applied to a fire. The two intrinsic advantages of such a system when compared to simple application of water are these. Water damage, which in some fires, especially in the sort of surroundings one might expect on a luxury liner, exceeds fire damage in monetary terms, is minimised by the use of the water mist rather than bulk liquid water. Also the small size of the droplets from such an extinguisher makes for rapid evaporation to water vapour, leading to a reduction in the partial pressure of oxygen close to the fire. This vitiating effect supplements the primary action of the mist, which is of course cooling. On QM2 there are 10 000 outlets ("heads") for water spray admittance and reticulation is by means of stainless steel piping of total length 58 km.

Quezon City, hotel fire

There was one death and about 20 injuries as a result of this fire at the Great Eastern Hotel, in Quezon City, the Philippines, in 2007. It has been reported that the hotel was not sprinklered.

Quint

The term is used to describe a piece of fire fighting apparatus. On a strict etymological basis (L. *quinque* = five) it means such an apparatus having five facilities: **vehicle-mounted pump**, water tank, hose, **ground ladders** and aerial ladders. A quint may or may not have a **ladder bucket**, as aerial ladders are sometimes made without such. One with a ladder bucket might sometimes be referred to as a quint/platform to distinguish it from one without. Current major manufacturers of quints include **Sutphen**.

Rack Hose™

Single jacket hose for use in buildings and plant. The lining is made of thermoplastic polyurethane (TPU) and the jacket of woven polyester. It lays flat in storage. It weighs about half as much as the corresponding product for brigade use.

Raffles Place, Singapore

Scene in the early 1970s of a department store fire. The store had been founded as far back as 1858 and by the time of the fire had a number of branches in the region and a franchise to sell Marks and Spencer products. In November 1972 the store was destroyed by fire with the loss of nine lives.

Rate-of-rise (ROR) heat detector

One which in addition to whatever heat sensing device it uses has a thermocouple, suitably compensated internally, so that the temperature history is retrospectively known. This is potentially useful in follow-up. A detector working according to photoacoustic infrared sensing might also be ROR.

Red phosphorus

Elemental phosphorus in this allotropic form is a good flame retardant for plastics. Like other phosphorus preparations so used it acts by denuding a polymer of the oxygen in its structure when heated, leaving a char which is resistant to the action of atmospheric oxygen. Trade names for red phosphorus as a flame retardant include **Exolit** RP, Amgard® CHT, Doverguard® 9021 and Hostaflam® RP.

Redwood, CA

The scene in November 2007 of delivery of a **tractor-drawn aerial** manufactured by **Seagrave**. Its "extras" are limited to tool storage. In general a tractor-drawn aerial will not be set up as a **quint** as it will lack a water tank and a pump, although it might carry hose. A tractor-drawn aerial very similar to that at Redwood had a few months earlier been delivered by Seagrave to the Fire Department in Collingswood, NJ.

Reel-Lite 800

Booster fire hose, like Pro-Lite a product of the Key Fire Hose Corporation. The lining is made of EPDM (ethylene propylene diene monomer) and the jacket of polyester. There is an outer cover of **nitrile**. Pressure testing is to 800 psi.

Refrigerant 22 (R-22), behaviour of on heating

R-22 is chlorodifluoromethane, CHF_2Cl. Data sheets on this substance declare it to be non-flammable except with an extreme ignition source. The reason for this is that a simple oxidation equation such as was given for **HFC-32** cannot be written for the more heavily substituted R-22, the response to which on heating in air is partly decomposition. Hence there is a need for more than an "ignition source" in the conventional sense to initiate reaction. The overall reaction is something like that represented in the box below:

Phosgene (very poisonous)

$\Downarrow$

$$2CHF_2Cl + 0.5O_2 + H_2O \rightarrow COCl_2 + HCOF + 3HF$$

$\Uparrow$

Formyl fluoride

Phosgene formation is known to be a hazard when R-22 is heated in air. R-22 is one of the HCFC (hydrochlorofluorocarbon) refrigerants as is **Refrigerant R-123**. (*See also* **R-403a**.)

Refrigerant R-123

1,1-dichloro-2,2,2-trifluoroethane, $CHCl_2CF_3$. Not flammable except when supplied with sufficient heat to initiate decomposition, as with R-22.

Refrigerant R-1270

Simply propylene (IUPAC name propene), C_3H_6. Obviously powerfully flammable. Similarly, propane in its refrigerant role is known as R-290. LFLs are respectively 3.1% and 2.1%.

Refrigerant R-134a

Chemical name 1,1,1,2-tetrafluoroethane, CF_3CH_2F. It is seen as being non-flammable under any conditions relevant to its function as a refrigerant. In moist air under pressure it will in fact burn according to:

$$CF_3CH_2F + 1.5\ O_2 + H_2O \rightarrow 2CO_2 + 4HF$$

Refrigerant R-142b

A.k.a. HFC-142b, refrigerant material composed of 1-chloro-1,1-difluoroethane which, being only half substituted with halogens, is quite readily flammable. The equation is:

$$CF_2ClCH_3 + O_2 \rightarrow CO_2 + 2HF + HCl$$

Its lower flammable limit (LFL) is 8%. The refrigerant is sometimes used alone but is more commonly part of a blend. Such a blend is R-405a, which has 5.5% of R-142b; its other constituents include **Refrigerant 22** and over 40% of octafluorocyclobutane (a.k.a. R-C318), which is of course totally substituted. There is no record even of an **elevated temperature flame limit** for R-405a.

Refrigerant 403a (R-403a)

Refrigerant of nominal composition propane 5%, chlorodifluoromethane 75% and octafluoropropane 20%. Too lean in combustibles to have a conventional lower flammability limit, it has an **elevated temperature flame limit** at 60°C of 13%. R-403b has the same constituents in different proportions.

Reogard®

Newly developed (by Great Lakes Chemical Corporation) phosphorus-containing fire retardants displaying **intumescence**, obtainable in two forms: Reogard®1000 and Reogard®2000. Each is applied to polypropylene, being blended with the polymer in melted form. Reogard® in either form will function as an intumescent fire retardant at lower loadings (that is percentage weight of retardant in the polymer/retardant blend) than many other such retardants will.

Republica Cromagnon Club

Venue in Buenos Aires, Argentina where in December 2004 there was a fire which caused 194 deaths and over 700 non-fatal injuries. Emergency exits at the club were locked, and fire procedures had been violated in other ways, for example in the materials of which the building was composed. Most of the deaths were due to inhalation of smoke and of carbon dioxide. The female rest rooms doubled up as a crèche and babies therein were amongst the dead.

Rescue vehicle

This term is of course widely used in its general sense in descriptions of fire incidents, but it is also a moderately precise term for a fire truck which does not carry ladders or water and does not have a pump. Such a truck therefore has a back-up role in carrying accessories such as tools and sometimes illumination facilities supplied with electricity by a power take-of (PTO) generator which might also power a winch. It is also likely to contain **SCBA** and a **thermal imaging** camera. The contents of a rescue vehicle are, in the words of one fire truck manufacturer, "limited by the imagination". It might consist of shelves and compartments all accessed from the outside. It might on the other hand consist of an enclosed storage area behind the cab, in which case it is a "walk-in rescue vehicle". One such manufactured by American LaFrance has recently been delivered to Sunnyvale, CA. This has a length of 11.4 m and uses a Detroit Diesel 455 hp power unit. A rescue vehicle which is not "walk-in" is "walk-around" if there is a partition along the axis of the vehicle, meaning that contents have to be accessed from one side or the other. Continuing to take examples from American LaFrance, it has recently supplied a walk-around rescue vehicle to Rexford, NY and a "rescue vehicle", having neither enclosure nor partition, to Yonkers, NY. Each of these is a sizable truck: the former is 10.9 m long and has a 400 hp Cummins engine, whilst the latter is 11.3 m long and has a 515 hp Detroit Diesel unit. Four further points will be made. One is that it is standard practice for fire fighters to travel in the enclosed part of a walk-in rescue vehicle when going to and from fires. Another is that some builders of such equipment, including Pierce, use the term "non walk-in" rescue vehicle. Next, Pierce offer a rescue vehicle "walk-in" at the rear but "non walk-in" closer to the cab, called a Combination. Finally, a rescue vehicle of any of the types defined here is often simply referred to as "a rescue".

Residential Hotel Sprinkler Ordinance, San Francisco

This law, which came into force in 2001, requires installation of sprinkler systems totally compliant with local codes in hotels defined as such according to the Ordinance. It followed a number of fatal fires in SFO in hotels not fitted with sprinklers, including those at the St Charles Hotel, the Hartland Hotel and the **Delta Hotel**. Lawsuits have been filed against certain hotels in San Francisco which have failed to install sprinklers as required by the Ordinance. A "success story" for sprinklers occurred in approximately the same part of the US with the studio fire at **Monterey, CA** in 2007. (*See also* **Chicago Hilton**; **Fort Collins, CO**; **Virginia Beach**.)

Rhino II

Bumper **turret** for use with an ARFF vehicle, developed and manufactured by Crash Rescue Equipment Services in Dallas. Its distinctive features include positioning such that "overspray" on to the windshield of the vehicle is prevented. From the same manufacturer comes the Torrent, a roof turret for ARFF vehicles. This can rotate through 210° of a theoretical maximum of 360° in a horizontal plane. Such scope for turret movement reduces the need for time-wasting vehicle movement in positioning prior to extinguishant release.

Rhoads Opera House

This building in Boyertown, PA was destroyed by fire in January 1908. The fire occurred during a performance on the second floor of the building and started when a kerosene lamp was knocked over. The one fire escape was inaccessible because the closest exit to it was locked, and its positioning was in any case very unsatisfactory. The death toll was 171, and several nuclear families perished together. State laws appertaining to fire safety in buildings came into effect the following year.

Rhode Island, Public Protection Classification (PPC™) of

Rhode Island is the second most densely populated state of the US after New Jersey. Its PPC™ profile is in the box below. It is close to that of New Jersey given in a subsequent entry of this volume in having a modal classification of 4 and only one class 10 community.

State	Highest classification	Modal classification	Number of class 10 communities
RI	2 (three communities)	4 (28 communities)	1
WY	3 (seven communities)	6 (56 communities)	21

The least densely populated of the "lower 48" states is Wyoming, figures for which are also given. The differential is typical of trends in going from densely populated places to sparsely populated ones. This information for Hawaii is classified. (*See also* **Selected US states, Insurance Services Office classifications for.**)

Robotics, contribution to fire fighting by

Only a very limited contribution as we approach the end of the first decade of the 21st century. There was a setback when a robot developed for such a purpose in the UK was taken permanently out of service after three years of use on an exploratory basis. This was the Fire Spy, whose joint developers were the West Yorkshire Fire Service and JCB, a manufacturer of excavators. The robot's primary role was seen as being entry into a building on fire and removal of containers of flammable substances before they were affected by the fire. It was believed that a Fire Spy robot could perform this task at temperatures up to 800°C as well as bearing such accessories as a **thermal imaging** camera and a hose. Development costs had been a fairly modest £30000. The West Yorkshire Fire Service introduced the Fire Spy in June 1999 and scrapped it in July 2002. Difficulties included its unreliability where its path was on a slope rather than flat, and its inability to surmount objects such as cables. Trinity College in Hartford, CT has for over a decade had an annual fire fighting robot contest, in which any group or individual from anywhere in the world can compete.[30] A robot entered for the contest needs to be able to go inside a mock building, locate a fire there and put it out. This contest, which is of very wide scope, seems to dominate R&D into fire fighting robots. (*See also* **Sintef; Shenyang; Ubiko T2-4.**)

Rochford, UK

The scene in early 2007 of a fire at a tyre retail outlet, which eight fire engines attended. Nearby roads had to be closed because of the effect of the thick smoke on driving visibility, and fire fighters remained at the site after extinguishment in case burning redeveloped.

Rockwall, TX

An example of a community having risen on the **Public Protection Classification (PPC™)** scale, from class 6 when first assessed in 1988, to class 5 in 1992 to class 3 in 2005.

---

30   Details on http://www.trincoll.edu/events/robot/.

Rocky Mountain states, Insurance Services Office classifications for

A breakdown of **Public Protection Classification (PPC™)** for these states is given in the table below.[31]

State	Highest classification	Modal classification	Number of class 10 communities
WY	3 (seven communities)	6 (56 communities	21
CO	2 (six communities)	6 (158 communities)	11
MT	3 (23 communities)	9 (80 communities)	31
NM	2 (five communities)	9 (186 communities)	19
UT	2 (four communities)	6 (132 communities)	21

The modal classification is the value corresponding to the highest peak in the histogram. Many of the histograms have more than one peak.
(See also **California, Texas and New York, Insurance Services Office classifications for; Selected US states, Insurance Services Office classifications for.**)

Romeoville, IL, refinery fire in 1984

A fire at the Union Oil Company refinery in Romeoville in July 1984 caused 17 deaths. It began with escape of vapour through a crack in a vessel wall. Fire fighting was impeded by damage to the water reticulation system. A tank of LPG underwent a Boiling Liquid Expanding Vapour Explosion (BLEVE). It was only the third industrial fire in the history of the State to have claimed more than ten lives.

Rosenbauer R240

Vehicle-mounted pump for fire fighting, of centrifugal design and capable of discharging about 650 gpm at a pressure of 10 bar. It is one of a series of such pumps bearing the Rosenbauer name one of which – the R600 – has a discharge rate at 10 bar of approximately 1800 gpm.

Rotunda

Building at the University of Virginia designed by Thomas Jefferson, third President of the US. Completed in 1826, it was destroyed by fire in 1895. The most valuable single possession within the building at the time of the fire – a life-size marble carving of Jefferson – was rescued by volunteers

---

31　The author has been unable to find this information for Idaho.

from amongst the student body. The Rotunda was rebuilt over the next few years, the wooden dome of the original being replaced by a tiled dome with good fire resistance properties. In 1976 as part of the bicentennial observance, the Rotunda was restored in close compliance with the original design by Jefferson.

Route 999 nightclub

In Pattaya Thailand, the scene in 2006 of a fire in which eight persons were killed and over 50 seriously injured. Seven of the dead were employees and the eighth a contractor, whose activity might well in fact been the origin of the fire as he was carrying out a welding operation on the air conditioning system.

Royco Hydraulic Oil 756

Hydraulic oil manufactured by Total finding application in aviation. Its kinematic viscosity at 40°C is 15.7 centistokes, close to that of **Nuto H**15. The flash point of 105°C is, however, much lower that of many such fluids. Sister products Royco Hydraulic Oil 777, Royco Hydraulic Oil 782 and Royco Hydraulic Oil 783 have flash points of respectively 175, 235 and 110°C.

RT700A®

Moisture barrier material from the same manufacturer as **Crosstech®**. It is based on PTFE.

S

S2 air tankers

These were made by adapting Grumman aircraft, which the US Navy were selling off by the mid-1970s and were converted into air tankers initially on a lease arrangement with the Navy. There are many still in operation and they form a significant part of the aerial fire fighting fleet of the California Department of Forestry. A typical example can hold 1000 gallons of water, which of course is likely to contain foam. Those in service in Canada are called the Conair Firecat.

S707

The latest in a range of intumescent coatings from Nullifire in Coventry, England. It is water based, and can be applied to internal steel structures, protecting the steel from reaching failure temperatures for up to two hours. It can be applied to steel members before construction or after. It has been used at a new retail building in central Glasgow, Scotland, where its rapid application was an advantage, as retail building projects often have to follow a very tight schedule.

Saberjet™

Fire fighting nozzle manufactured by Akron® with the capability to produce a straight stream, fog or a **combination stream** depending on the setting. Unlike a **single gallonage nozzle** it does not operate at a particular pressure but at pressures in the range 50 to 100 psi, the delivery rate depending also on the orifice diameter, which is fixed once installed but selectable, from a range, on purchase. It has been used with **compressed air foams**.

Sabre

Water ladder from the UK manufacturer Dennis. Configurable to customer's requirements to a considerable degree, a typical example will have a water tank holding ≈ 450 gallons. A pump from the **Godiva World Series** will be incorporated. The Sabre is used by many UK fire authorities.

Safire

Proprietary intumescent material used in fire protection for applications similar to those of **Metacaulk® 1100**. Its ingredients include polyvinyl acetate, polyvinyl acrylate and natural latex.

St George, UT

In the US and the UK small fire departments are sometimes the first to use **PPV**, and their recorded experience will no doubt be valuable to larger fire departments considering PPV. An example is St George, UT, population 68 000. A fire apparatus of which the St George Fire Department took delivery in 2007 has two PPV fans.

St Lucia

This Caribbean island received in 2002 three fire appliances manufactured by **Angloco**. They were designed for the poor road conditions and steep terrain. Angloco appliances are also in service in Barbados.

Salang tunnel fire

Reliable information on this tragedy is not abundant. The Salang tunnel links northern and southern Afghanistan, providing a route through mountainous terrain. The fire there occurred in 1982, at which time Afghanistan was under Soviet occupation. The fire began with the collision of two heavy vehicles inside the tunnel; possibly one of them was a tanker carrying flammable liquid. The occupying army blocked off exit from the tunnel after the collision, so to fatalities due to the fire *per se* were added fatalities due to inhalation of carbon monoxide from vehicle engines. Estimates of the death toll vary between 1000 to 2000, the majority Afghan civilians, although undoubtedly some soldiers serving in the occupying army were also amongst the dead.

Salem, OR

Scene of a fire in June 2006 involving biodiesel. The fire occurred in a barn where a small business manufacturing and storing biodiesel for sale to rural consumers was based. Biodiesel had been stored in small containers and there was also methanol for use as a reagent. On being sent to the fire, the fire fighters thought that on arrival they would simply be extinguishing a blaze at a barn. Once there they found a fire much more powerful than any which the building alone could have caused and were informed at that stage of the contents of the enclosure.

Sandberg bluegrass

Botanical name *Poa secunda*, a **firewise plant** owing that property to a low fire load as with **western wheatgrass**.

Santa Clara, CA, five-alarm fire in 2004

On 9 July 2004 there was a fire at an apartment building in Santa Clara that was assigned 5 on the **alarm number** scale. It occurred in the small hours. One occupant of the ground-level apartment where the fire began was killed. Second-floor residents, whose smoke alarms were activated, called the Fire Department.

Sao Paulo Airport, Brazil

The scene of an accident claiming about 200 lives when an airliner on a domestic route crashed on landing and caught fire. Burning continued for many hours. It is probable that initial combustion was of the fuel, which of course is kerosene, obtained from crude oil in the boiling range between naphtha and diesel. A fireball originating from this was in fact observed. Once there is an initial fire there is a good deal of material for continuation of combustion, including cabin fittings. An acrid black smoke resulted from the burning of these.

Sargent cypress

Botanical name *Cupressus sargentii*, this species of tree, classified as a **highly flammable tree**, occurs along the coast from northern to mid California. This is also the region of occurrence of the Gowen cypress *Cupressus goveniana*: the two species have many similarities and are sometimes confused. Gowen cypress does not appear in lists of highly flammable trees.

SAS Nagar

Location in the Punjab, the scene in 1999 of a fire involving ammonia. The ammonia was in use as a refrigerant at an ice factory. There were no injuries and extinguishment was prompt. Even so, the fire fighters reported that parked vehicles close to the factory had impeded their operations.

Scandinavian Star

Car ferry which in April 1990 was the scene of a fire which claimed 159 lives. The vessel had been built in 1971 and, under a succession of owners and names, had previously been in service as a casino ship in places including the Mediterranean and the Florida coast. In the new role of a car ferry she began operating between Norway and Denmark in 1990.

On the day of the tragedy two fires broke out on the vessel one of which, it was concluded, was the work of an arsonist. A catalogue of difficulties relevant to the management of the fire and responsible for its escalation became evident in the investigation, one of which was that the training period of the crew had been significantly shorter than is usual for a vessel the size of the *Scandinavian Star*. Doors which ought to have closed automatically to prevent spread either had to be closed manually or could not be closed at all. The ventilation system, which might have kept the smoke sufficiently dilute to reduce effects of smoke toxicity, was at the instruction of the captain turned off and this enabled smoke to enter cabins occupied by passengers. Part of the interior of the area of the vessel where the fire began contained panels made from a substance known to produce hydrogen cyanide and carbon monoxide on combustion.

The vessel was not totally destroyed by the 1990 fire and was eventually repaired at a shipyard in Italy. Having had a succession of further owners and names, it was taken to the ship breaking facility at Alang, India in 2004.

Scania

This manufacturer of fire trucks offers **water tenders** with a range of water containment capacities of up to 4000 litre (> 1000 US gallons). Foam in quantities of up to 500 kg can be carried and, as is often the case with a water tender, there is a further role in carrying ladders, rope and the like.

S-Cap

Escape mask manufactured by MSA. It provides protection from carbon monoxide for at least 15 minutes by reason of a **Hopcalite** catalyst, as well as removing certain other gases, including sulphur dioxide, and filtering out particulate. The external parts are bright yellow, to aid in visibility, and the transparent part of the mask in front of the eyes provides for a wide field of vision.

SCBA

Self-contained breathing apparatus. Such a device for fire fighting[32] has a cylinder of compressed air sufficient, depending on cylinder size, for up to one hour of sustained use. A supply rate of about 300 litres per minute is required. The "positive pressure" SBCA is the type now most widely used,

---

32 Not all **SCBA** is for fire fighting. Some are for entry into areas with a hazardous atmosphere where there is no fire at all. These will often conform to NIOSH but not to NFPA requirements. Sometimes a particular make and design of SCBA is available either in NIOSH-compliant form or, at greater expense, in NIOSH and NFPA-compliant form.

in which the enclosure between the inside of the mask and the fire fighter's face is maintained at a pressure slightly above that of the atmosphere even when the fire fighter inhales. This ensures that if the mask is not a perfect fit the result is harmless leakage of clean air and not ingress of contaminated air from the outside. The cylinder and its attachments are worn on a harness and this is a significant factor in fire fighter fatigue. Aluminium is a common choice of cylinder material because of its low density, although materials have been developed for this purpose which are less dense still. In the US cylinders for SCBA are designed for one of two internal pressures: 2216 psi or 4500 psi. A cylinder once depleted can be refilled at the scene of a fire either from larger containers by means of a **cascade system** or from ambient air by means of a compressor. Parts of the mask which fit above or behind the head will include materials such as **Nomex**®. Commonly (e.g. in the case of the **Panther**™) the face piece of a SCBA will have a speaking diaphragm, enabling the wearer to send an oral communication.

Schiphol Airport, fire at

In October 2005, a detention centre for illegal migrants at Schiphol Airport in Amsterdam was the scene of a major fire. At the time of the fire the detention centre had 350 inmates, 11 of whom were killed, whilst a further 15 were hospitalised. The detention centre is close to one of the runways. A Libyan national who had been a detainee at the centre at the time of the fire was arrested and charged with having started it. He had asked for and received a cigarette lighter shortly before the fire. He was due to have been deported on the following day.

Over the period 2003–2005 an area of Schiphol airport had received a coating of the intumescent substance **Chartek 7**.

Schomberg Plaza

High-rise apartment block in the Harlem district of NYC, the scene of a fire in 1987 in which seven people were killed. The fire began at a compactor in a garbage chute. During the enquiry it transpired that the sprinkler system at Schomberg Plaza had not been working for some time prior to the fire and that throughout the 35-storey building fire protection was poor. Once the FDNY arrived they put out the fire at the garbage chute but did not turn their attention to parts of the block to which the fire had spread, to the endangerment of persons, until 16 minutes after arrival. Addressing a Grand Jury, the Fire Commissioner stated *inter alia* that since the opening of the apartment block 13 years earlier the Fire Department had responded 33 times to incidents originating at its garbage compactor.

Schools, arson attacks on

On average, each week sees 20 arson attacks on schools in the UK, a third of which occur whilst the school is occupied by children and teachers. Commonly, arsonists are minors who are former pupils of the target school or have a brother or sister there. As such they will not have access to accelerants or knowledge of how to use them, and they tend to make for such things as ignitable rubbish in the playground area. One mitigation measure therefore is the siting of waste receptacles such as skips well away from the building, behind locked gates if possible. (*See also* **Castleford High School; Copleston High School; London schools, (absence of) sprinkler systems; Lord Williams's School.**)

Scotsdale, AZ

Data on the efficacy of sprinkler systems was obtained from a study of fires in Scotsdale over a ten-year period. The study related not only to whether a sprinkler system was fitted at all but to the number of **sprinkler heads** in those of the buildings which had been so equipped. Over the period of the study there were 574 fires in commercial premises: in only 65 of them was a sprinkler system fitted. In 59 of those 65 the fire was brought under control by a system comprising one or two sprinkler heads. In the others more sprinkler heads than two were activated, up to a maximum of five.

Scotty Five-Litre Foam Kit

Example of an **eductor**, the entire "kit" comprising that and a container of foam concentrate to which it is attached. It is suitable for use with **AFFF** proportioning and will discharge 50 gpm of water having been charged with foam. Its construction materials are non-metallic. It is the largest in a range of eductors of that particular design and configuration: the smallest delivers a mere 3 g.p.m. and can be used with garden hose. The same manufacturer makes eductors for heavier duty use (up to 95 g.p.m.). Known as the Scotty 4046 series, these are intended for use with an **aspirated foam nozzle**.

Seacliffe Mental Hospital

Thirty-seven patients died as a result of a fire in this building on New Zealand's South Island in December 1942. It housed 500 patients and 50 resident staff and was the largest building of its kind in the country. The nucleus of the building was stone, but a two-storey wooden extension had been made and this was where the fire occurred. Patients were locked in their rooms and there was no duty nurse in that part of the hospital. Destruction of that part of the hospital was rapid, with heavy loss of life as noted. A Commission of Inquiry noted *inter alia* that the part of the hospital where the fire was had no fire alarms. The cause of the fire was never discovered.

Seagrave

Manufacturer of fire trucks, based in Clintonville, WI (though having manufacturing facilities in other parts of the US including South Carolina) and a major US manufacturer of such appliances. Its products include the widely used Marauder II pumper, the structure of which is stainless steel. Its "aerials" include the Meanstick, which has a 75 foot ladder and the capability for below horizontal operation. These share structural features with the Marauder, as does the sister product the TowerMax, the ladder reach of which is 104 feet. The company's range also extends to tankers in which bodies can be stainless steel or aluminium and specifications including tank capacity are to customer's requirements. Seagrave also manufacture a range of **tractor-drawn aerials** which use the Marauder as a basis for the cab. One of the many consequences of 9/11 was heavy loss of appliances owned by the NYFD. Seagrave received many of the orders for replacements and met these on schedule without defaulting on existing orders or refusing new orders. One of the appliances from Seagrave – a pumper – was supplied at no cost. Pierce also donated an appliance as did **Crimson**, Rosenbauer and American LaFrance. Part of the cost of the Seagrave appliances was met by the citizens of Akron, OH who co-ordinated a fund raising campaign which realised US$1.4 million.

Selected countries, fire death rates in

During the first few years of the 21st century, the USA had an annual death rate of ≈12 deaths per million population. Australia is significantly better at ≈7 deaths per million population, a figure which also applies to Italy and the Netherlands. New Zealand and the UK have rates of ≈10 deaths per million population. Denmark, Greece and the Republic of Ireland have rates of ≈16 deaths per million population; Japan is a little higher than this at ≈17 deaths per million population. Finland and Hungary each have death rates marginally exceeding 20 deaths per million population. The value for Canada is year by year quite close to that for the USA. (See *also* **Fire services, costs of in selected countries**.)

Selected US states, distribution of Insurance Services Office classifications for

This information for ten states is in the table below. The reader might care to reflect on the following. First, many states have 9 as a modal value classification. Secondly, New Jersey, Connecticut and Massachusetts appear to be superior in PPC™ terms to New York, corresponding data for which are given in another entry. Thirdly, there are very many class 10 communities. Properly understood this means no more than that no relief on fire insurance premiums can in those communities be expected on the grounds of the level of "Public Protection".

State	Highest classification	Modal classification	Number of class 10 communities
NJ[33]	1 (one community)	4 (240 communities)	1
OK	1 (one community)	9 (733 communities)	133
KY	2 (12 communities)	9 (342 communities)	4
CT	1 (two communities)	5 (71 communities)	4
AK	2 (two communities)	8 (23 communities)[34]	7
IA	3 (25 communities)	9 (863 communities)	36
MA	1 (one community)	4 (115 communities)	9
ND	3 (three communities)	9 (424 communities)	90
KS	2 (six communities)	9 (378 communities)	51

(See also **Rhode Island, Public Protection Classification (PPC™) of**; **Rocky Mountain states, Insurance Services Office classifications for**.)

Sembewang Shipyard

This shipyard in Singapore, which occupies the site of a former British naval base, was the scene of a fire in September 2007 in which there were two fatalities. The victims had been working on a vessel brought into the shipyard for repair. At Sembewang many types of vessel, including oil tankers, are received for repair or modification. Over the years the shipyard has had a highly creditable safety record having in successive audits by Det Norske Veritas (DNV) received ratings of 7, 8 and 9 on a scale which goes to 10. The recent accident will provide input for risk assessment.

SenseTronic® T-229/2

Flame detector responding both to UV and to IR, to wavelengths in the 185 to 235 nm range in the case of the former and at 4.4 μm in the case of the latter. (See also **UV/IR flame detector**.)

---

33   The most densely populated state of the US

34   There is a class 8b, which signifies eligibility for class 8 apart from in the matter of water supply and is therefore better than class 9. In this figure 8 and 8b have been lumped together by the author

Shanghai, fire in 2005

A fire at a hostel for migrant workers was the scene of a fire in 2005 when 10 people died and 19 more were injured. The building had previously been a factory and it is understood that no approval for it to be used as a residence had been obtained and that fire regulations had not been complied with. Shanghai is a disorderly city of population over 15 million and the fire authorities work in a difficult milieu. One difficulty is poor access of fire appliances to crowded parts of the city. Although there are too few fire appliances per unit area of population, many of those in service are modern and well equipped and there are also helicopters. There are encouraging statistics indicating fewer deaths from fires in Shanghai in the last year than previously. There is also evidence that the city is pro-active in such matters, for example by observance of an annual "fire fighting day".

Sharon Hill, PA

The scene in 2007 of a fire which claimed the life of a 19-year old fire fighter serving with a local **Good Will Fire Company**. He was one of three fire fighters trapped through collapse of a wall; the other two were hospitalised.

Shenyang

A major fire in this Chinese city incurring very heavy losses led to the development of a robotically operated fire truck. The entire truck can be moved in response to directions from a human fire fighter up to 500 m away, and the water and foam functions of the truck can be brought into use similarly.

Shinagawa Thermal Power Plant

Facility of the Tokyo Electric Power Company, using imported liquefied natural gas as fuel. Fire protection there is by **Nippon Dry Chemicals** who have installed extinguishing devices using **Inergen** as well as indoor and outdoor foam extinguishment systems and dry chemical systems.

Ship's engine, use of to drive fire pumps

Usually fire pumps on a vessel, upon the capability of which the FiFi classification depends, are driven by pump engines. For example, the *Bulldog* has two 900 hp diesel engines to serve its water pumps as has the *Lynne Moran*. In some of the most recently built tugs the pumps are driven by the tug's own engine, making the vessel lighter because of the absence of fire pump engines. Tugs of this sort have been awarded **FiFi I** classification and there are plans to introduce such tugs at the **Neptune LNG terminal**. The *Bulldog* is also in fact used in LNG handling.

Ship's engine room, a recent issue in the fire protection of

The following interesting and important point was made at a meeting in 2006 of the Sub-Committee of Fire Protection of the Intergovernmental Maritime Organisation (IMO, formerly IMCO), when criteria for insulation performance in the protection of ships' engine rooms were under discussion. Tests by means of which an insulator is assessed were developed a century ago when the flammable contents of a ship were primarily the wood in its structure. Now, the most serious hazard is liquid fuel and lubricating oil. It therefore appears that tests appertaining to cellulosic fires are being used in protection against hydrocarbon fires, which have much higher heat-release rates. This revelation formed the basis of a case for a move from conventional insulators, such as mineral wools, with a merely insulating function to materials capable of **intumescence**.

Showa Oil Niigata Refinery

There was a major earthquake in Niigata in 1964 and at the Showa (Shell) Refinery there crude oil ignited. It is known that **boilover** occurred and contributed to the involvement of other tanks of crude oil in the fire. Niigata receives much snow and the **deep snow fire hydrant** is in use there. (*See also* **Toa Oil**.)

Sikeston, MO

A second-hand fire truck recently bought by the Sikeston Rural Fire Protection District is expected to lower insurance premiums for home owners in the district. The truck is a 2700 gallon water tanker sourced in Oregon and carried by trailer to its new base in Missouri.

Sikorsky S61

Helicopter whose uses include fire fighting. A water uptake device ("snorkel") enables a quantity of 1000 gallons of water to be taken from a river or lake in less than one minute. In size and performance the S61 falls between the **Erickson S-64** (2650 gallons) and the **Huey** (360 gallons).

Silica, as a fire retardant

This very abundant mineral has, in a number of forms, found application as a fire retardant for polymers including polypropylene and polymethyl methacrylate (PMMA, a.k.a. Perspex). Its action, like that of many other retardants for polymers, is to inhibit pyrolysis. It is also a component of some fire retardants which show **intumescence**, when its surface area is an important factor, so fine grinding of the silica additive is beneficial.

Silicaflex™

Fire-resistant material composed of over 95% silica on a weight basis, suitable for applications including fire blankets. The balance of the weight is accounted for by a proprietary hydrocarbon material which enhances the flexibility and abrasion resistance of the material. In addition to fire blankets, fire-resistant sleeves for vulnerable parts of machinery can be made from Silicaflex™ and whole walls can be protected by it, in "curtain" form or by means of an adhesive coating at one side (in which case the material is known as Silicaflex™AB). Silicaflex™ will withstand for sustained periods without loss of function temperatures approaching 1000°C and for shorter periods temperatures of 1650°C.

Silicon carbide

Semiconductor suitable for gas detectors, commonly those for hydrogen and hydrocarbons. Greater distinction between one hydrocarbon and another is possible with silicon carbide than with **tin(IV) oxide**.

Silicon photocell

A photovoltaic device, that is it generates electrical current in response to incident electromagnetic radiation. It responds most to radiation encompassing the high wavelength end of visible and the low wavelength end of infrared, hence its suitability for use in a **spark detector**.

Silver buffaloberry

Botanical name *Shepherdia argentea*. This shrub occurs widely in the US and Canada, from British Columbia and Alberta down to California and as far east as Iowa. It is used as a **firewise plant** in states including New Mexico.

Sintef

This Norwegian company has developed a hose comprising short segments, enabling it to be orientated robotically instead of by a fire fighter. In addition to movement of the segments relative to each other, making for flexibility, up and down movement of the hose is possible.

Sirocco 9303

Type of fire fighting hose, commonly stored on a reel for use in buildings. Both lining and outer jacket are made of a thermoplastic elastomer (TPE) and there is reinforcement by woven textile. Depending on diameter, it can operate at pressures in the range 90 to 225 psi. **SBR** is used similarly to TPE in the fabrication of hose for interior use.

Skandi Yare

Offshore supply vessel currently on charter to the Brazilian company Petrobras. It is to be retrofitted, as part of an extended charter arrangement, with equipment to bring it to **FiFi II** standard. Pumps and **monitors** are available which are portable. This is true of the **FiFi I** equipment currently being fitted to two vessels operated by the Norwegian owner/operator Deep Sea Supply, which has heavy involvement in the Norwegian sector of the North Sea.

Skum

Pronounced "skoom", Swedish manufacturer of fire protection products including foam generators. Amongst its products is the HotFoam® system. This is conventional in that is uses water and foam concentrate combined by means of a proportioner but different from other such devices in that the foam before application is aerated, ambient air (or indeed smoke from the fire itself) being used for this. Its applications include ships, and it is seen as a replacement for **Halon 1301**.

Slick™

Nomex® fabric in a form such that it is suitable as a **moisture barrier**. It is a product of Lion Apparel whose HQ are in Dayton OH. Lions's other products include **Isodri®**.

Slopover (1)

Effect similar to **boilover**. Slopover occurs when water from fire extinguishment enters a burning liquid and descends, to be evaporated. Its subsequent rise through the liquid exacerbates the fire by destabilising the bulk fluid as with boilover. Slopover is less vigorous than boilover, and one does not expect the hydrocarbon released to ascend but to froth over the side of the tank making containment with a suitable foam agent possible.

Slopover (2)

In forest fires, progression of the burning front beyond either a natural boundary or one constructed by means of fire breaks, possibly marking the transition from "forest fire" to "wildfire". A.k.a. "breakover".

Smit Kamara

Anchor handling, tug and supply vessel in service in the Netherlands. Sister vessels are the *Smit Komodo*, which is on charter in Egypt, and *Smit*

Nicobar, which is currently at Sakhalin. Each has fire fighting capability at
FiFi I level.

Smokebloc®

Flame retardant for PVC manufactured by Great Lakes which, unlike
Ongard®, contains antimony. It also contains zinc and, in some forms,
silicon and/or boron. Though developed by Great Lakes for PVC it does find
application to other polymers, including polyurethane, and (like **Firemaster®
552**, also a Great Lakes product) to adhesives.

Smoke detector, in pre-alarm mode

For a smoke detector to be activated needlessly is at least a nuisance and
possibly a cause of financial loss. Reasons for such activation include entry
of harmless dust. Some optical alarms are set up so as to be capable of
pre-alarm status. One such, in which the signal from the diode is in analogue
form, sets 45 as pre-alarm value and 55 as alarm value, finally sounding
an alert on attainment of 55 but in the event of the maximum being in the
45 to 55 range disregards the signal. This reduces the number of needless
alarms. The same detector reads 25 in the arbitrary units in the absence
of smoke or other particles, and a value of 3 denotes failure.

Smoke detectors and heat detectors, co-existence of

The two have their relative merits but in many applications are equivalent
in effectiveness. Where there is one of each type protecting the same
area, it is recommended that they be interconnected so that if one goes
off so does the other. The reason for this practice is easy to understand:
the probability that both will fail is smaller than the probability that one will.
This is explained in the boxed area below, with arbitrary but reasonable
figures for the reliabilities.

Let each detector be 98% reliable, i.e. fractional reliability = 0.98

Probability that smoke detector will fail = 0.02

Probability that heat detector will fail = 0.02

Probability that one or other will fail = 0.04

Probability that both will fail = $0.02^2 = 2 \times 10^{-4}$

By interconnecting the two, the probability of there being no
response lowered by the factor:

0.04/0.0002 = <u>200</u>

Smokejumper

Fixed-wing aircraft from which fire fighters enter a forest fire by parachute. An obvious danger from this means of getting fighters to a fire is that they can be trapped without any vehicular means of escape. However, usually a decision to parachute fire fighters is taken only in the early stages of a fire when its extent is not huge and fire fighters can escape on foot in a direction other than that of propagation. Notwithstanding the danger from isolation, deaths from use of smokejumpers over a 70-year period have been comparatively few. An exception is the fire at **Mann Gulch, MT** in 1949 in which 13 fire fighters, having been parachuted into a forest fire, were killed. In the US there are smokejumper bases in Montana, Alaska, Idaho, Washington, California and Oregon.

Smoke suppressant

In a polymer of fabric material a smoke suppressant works by catalysing combustion, and this is at first a little counterintuitive. However, flammable degradation products in a flame can be made, under the influence of a smoke suppressant, to burn completely to gases instead of going along a path to smoke production commencing with polynuclear aromatic formation. These precursors to smoke formation, which accompany smoke particles themselves by adhesion to the surface, are carcinogenic so their elimination is a spin-off benefit from smoke suppression. **Molybdenum trioxide** is a common example of a smoke suppressant.

Snozzle®

Water tower manufactured by Crash Rescue Equipment Services in Dallas. It also conforms to the definition herein of a **hydraulic ladder** but differs from many such in that it is usable as an **attack apparatus** when connected to a **water tender** via a pumper, the intended *modus operandi* of the appliance.

Snozzle™

Device whereby the fuselage of an aircraft can be pierced in order that fire extinguishing agents (water, foam) can be injected. It has to be some safe distance from the human occupants of the **ARFF** incorporating the device and also has to be targeted accurately. This is achieved by means of support on a moveable structure which works similarly to a **hydraulic ladder**.

Soaptree yucca

Botanical name *Yucca elata*. This **firewise plant** is indigenous to parts of the USA (Texas, New Mexico, Arizona and Colorado) and to northern Mexico. It grows up to five feet in height.

Sodium antimonite

Chemical formula $NaSbO_3$. It is used as a flame retardant in the presence of a halogen-containing substance such as **brominated polystyrene**, the two acting synergistically. Antimony trichloride can act similarly.

Sodium bicarbonate

Dry extinguishing agent which works by production of carbon dioxide. Commercial preparations available for such use include BC25, which contains 25% sodium bicarbonate at particle sizes < 350 μm. In the performance of sodium bicarbonate and **potassium bicarbonate** in fire extinguishment, particle size is important.

Sodium lauryl sulphate

Chemical formula $C_{12}H_{25}O_4SNa$, widely used as an ingredient of fire suppression foams. When foams are designed for particular applications, an important property is the extent of expansion. An enclosure fire might well require a foam which expands strongly, whilst in an out-of-doors fire, for example involving spilt hydrocarbon liquid, a foam displaying only a low degree of expansion is preferable. The expansion properties of a particular foam are within limits controllable by adjusting the amount of sodium lauryl sulphate present. There are numerous "recipes", some of them patented, for fire extinguishing foams and many of them include either sodium lauryl sulphate or ammonium lauryl sulphate. Other ingredients might include a water softener, a buffer and an agent for adjusting the viscosity.

Sofa Super Store

In Columbia, SC, the scene in July 2007 of a fire which killed nine fire fighters, the worst such loss in the US since 9/11. Serious deficiencies were identified, including padlocked doors and the absence of a previously rehearsed evacuation drill for store employees. The store owners were required to pay fines amounting to US$13 000.

Solid state detectors, use of in fire protection

Solid state detectors and analysers, examples of which are multitudinous in today's world, work on the principle of ion formation from the material being analysed and an electrical signal from the ions when they impinge on a semiconductor. For example, in hotel and restaurant kitchens where propane is used as fuel, a detector for it will release photons which selectively ionise propane. Ions so formed contact a metallic oxide semiconductor (MOS) and create a signal which can be the basis of an emergency response. The technique is also used in examining fire remains for accelerants where arson is suspected. Sometimes a particular detector is available with either solid state or electrochemical sensing according to customer choice as with the Z Gard®.

Solomon Islands, fire services in

This nation, which forms part of the Pacific region known as Melanesia, is needing assistance from larger countries in many areas of its activity, including fire fighting. Help with that is being received from New South Wales (NSW), in the form of secondment of a NSW fire officer of Commander rank to the Solomons for a two-year period. The officer so selected was previously involved in rural rather than urban fire fighting, his "patch" having been the Blue Mountains to Sydney's west and the plains beyond them. Five fire appliances have been taken from Australia to the Solomons, having been carefully selected in terms of suitability for the conditions there. In the Solomons the fire service is formally a branch of the police. There are police from several countries, including Australia, in the Solomons at the present time and the NSW Fire Commander will work alongside them in shaping a future fire service for the country.

Solomon's Island, MD

The scene in March 2006 of a fire involving two adjoining restaurants. The efforts of the local fire authorities in preventing spread to other nearby buildings were supplemented by those of a local tug operator. Two tugs having suitably positioned themselves sprayed water on to the fire. Each tug had fire fighting capability at **FiFi I** level. There were no deaths or injuries. The damage bill of US$5 million would have been much higher had it not been for the participation of the tugs.

Sonoma County Airport

In Santa Rosa, CA, the recipient in August 2006 of a newly built **Jaguar Colet** K/15S **ARFF** at a cost of US$620 000. The K/15 had by then undergone some significant advances on the earlier versions delivered to places including **Hartsfield International Airport** about a decade earlier and that supplied to Sonoma County is a "Generation 4" model. When holding its payload of 1500 gallons of water, it weighs 14.5 tonnes, a low weight for such an appliance attributable in part to the aluminium panels. It also has a somewhat smaller (diesel) engine than some K/15s do, actually a 500 hp Cummins unit. Nevertheless acceleration – an important attribute of such a vehicle – is lively, 50 mph being achievable in 20 seconds or less.

Sound alert localiser

An **automatic distress signal unit (ADSU)** releases sound of a particular frequency, say about 3 kHz (as with the **Diktron DSX**). The human hearing faculty can have difficulty locating the source of sound at a single frequency, but can very readily and reliably locate the source of sound encompassing the entire audible range of frequencies (broadband sound, a.k.a. "white sound"). Consequently a device has been developed whereby after a short preset time of single frequency sound, emission broadband sound replaces it. Termed the sound alert localiser and developed in the UK, it awaits commercial implementation.

Sounders, for fire alarms

These can be integral to the alarm, in which case the decibel level will have been preset, or external and connected to it in which case the decibel level might well be adjustable (as with the **Banshee™**). Such a sounder will seldom release noise at a fixed frequency but with a frequency sweep, typically 800 to 950 Hz over a sweep time of a few seconds.

South Korea, fatal construction site fire

There were 40 deaths resulting from a fire at an almost complete warehouse building in **Inchon** early in 2008. Polyurethane had been installed at the building, which also contained refrigerant plant.

South Lake Tahoe, CA

The scene in June 2007 of forest fires affecting an area of over 3000 acres and destroying hundreds of buildings, most of them residential. It is believed that an illegally lit camp fire was the origin of the disaster. There were no deaths, but a few minor injuries amongst the fire fighters. Helicopters for fire fighting were brought from neighbouring areas, there being none such based at Lake Tahoe. **Hydromulching** took place after the fires, the operation taking 15 days to complete.

Spark detector

Such a device might be required where, for example, there is bulk handling of solids and the possibility of spark generation by friction. Associated temperatures are around 700°C and the detector, often a **silicon photocell**, needs to respond to radiation in the approximate wavelength range 1.5 to 3 μm. The signal from the detector will initiate suitable emergency measures, usually operation of a water spray. Where a spark detector is installed in a duct, as in solids handling it often will be, there will be no false alarms due to daylight or artificial light. If it is not so installed, it might be necessary to position it in a dark location to provide the necessary contrast.

SPH 100

State-of-the-art **quint** built by **Sutphen**. One such was delivered in May 2007 to Genoa Township FD, OH. In addition to its aerial ladders it has a 2000 gpm pump manufactured by Hale and a 300 gallon water tank.

Spiderwort

Botanical name *Tradescantia virginiana*, flower occurring widely in parts of North America and recognised as a **firewise plant**. The same characteristic is shared by *Tradescantia occidentalis* a.k.a. western spiderwort.

Spring Lake, NJ

Scene of a **Good Will Fire Company**, the assets of which include a **Sutphen** pumper which has been in service since 1985 and can deliver 1500 gallons per minute. It also has a 1930 pumper, manufactured by **Seagrave**. Though now primarily of historical interest, this pumper has in fact been maintained in operational condition by the Company.

Sprinkler head

This will receive water under pressure, and a valve within it will open to release the water in response to a temperature-sensitive device such as a **frangible bulb, fusible link** or a bimetallic disc. On water exit from the sprinkler head, pressure energy is converted to kinetic energy, but if as is usually the case the water is directed downwards, this will have only a small dispersion effect as the movement is in the same direction as gravity. The water is therefore passed against a deflector, which gives its velocity a component orthogonal to gravity and makes for wider distribution. A spray pattern is in this way established. The deflector is usually easily visible when a sprinkler head is examined, comprising a horizontal surface with a toothed boundary. Use of the temperature-sensitive initiating device means that not all of the sprinklers protecting a particular part of a building will open in response to a small, localised fire, but that sprinklers will open as necessary if the fire is able to propagate. That of course prevents non-thermal damage due to water. If this control is removed so that all the heads open at the same time, the system is not a sprinkler system but a *deluge* system. Such are common for example at offshore platforms, where non-thermal damage is of no consequence and seawater is available. (*See also* **Scotsdale, AZ**.)

Sprinkler pipes, materials for

Galvanised steel is a common choice. This has a moderate corrosion propensity in this application but has the advantage of not losing its ability to function at higher temperatures. This means that sprinkling can continue some time into fire development. By contrast **CPVC** shows no chemical deterioration due to water contact at all, but will start to lose mechanical strength at about 70°C. Polybutylene, the other non-metallic substance used for sprinkler pipes, is only reliable up to about 50°C. Copper has also found application to sprinklers and, depending on the precise requirements, is by no means always the most expensive option.

Sprinklers, frequency of accidental operation of

Factory Mutual give this figure, as it relates to manufacturing defects, as 1 in 16 million sprinklers per year, a figure which is examined in the boxed area below.

Frequency with which one in 16×10^6 sprinklers will go off = 1 $year^{-1}$

Frequency with which one particular sprinkler will go off

= $[1/(16 \times 10^6)]$ $year^{-1}$

= 6.25×10^{-8} per year

It would on the basis of the above be reasonable to express a frequency of the order of 10^{-7} $year^{-1}$ as a reliability index of one sprinkler.

Sprinklers, insurance benefits from for residential buildings

Sprinklers in private homes[35] in the US is a "growing culture" at the present time. There are ordinances relating to sprinklers in homes only in quite a small number of communities across the entire country and these relate only to newly built homes. Insurance premium reductions expected from the installation of a suitable sprinkler system are in the range 5 to 15%. More precise statistics than are currently available on the benefits of sprinklers in private residences are awaited. The most pro-active state in terms of sprinklered homes appears to be Illinois, where additionally there has been the "sprinklering" of the **Tribune Building** and of the dorms at **Elmhurst College**. (*See also* **Indian Hills, KY**.)

Sprinkler systems, antifreeze for

Under many climatic conditions sprinkler water will require an antifreeze. Those endorsed by the NFPA include glycerine and propylene glycol.

---

35 "Home" here means single-family home, quite simply a house. The requirements of "condos", apartments and home units where the populations are denser are different.

Methanol has also been so used. Where the pipe work for the sprinkler system is metal, attack by an antifreeze agent is not expected. When **CPVC** is used as the pipe material, however, only glycerine is a suitable antifreeze.

Sprinkler systems, for homes, cost of (USA)

Early in 2008 an approved sprinkler system added US$1 to US$1.50 per square foot of floor area to the cost of building a new house. Retrofitting an existing home with sprinklers tends to be more expensive and >US$5 per square foot is not uncommon.

Sprinkler systems, recent case histories in the UK

Argus Fire, themselves suppliers and installers of sprinkler systems, have in their promotional literature reported the following recent fires in the UK in which the consequences were much less serious than they would have been in the absence of sprinklers.

Region of the UK	Circumstances of the fire	Sprinkler performance
South Wales	Arson in a store room.	One **sprinkler head** only came into operation and the fire was extinguished. Store open for business on the next working day.
South Wales	Fire in plastic storage crates in a service road in a shopping centre.	Sprinkler system brought into operation by the heat. No spread of the fire into the shops.
Scotland	Explosion in a pharmaceutical plant.	Sprinkler system prevented fire from developing as a result of the explosion.
East Anglia	Fire at a paint-manufacturing plant.	Fire fighters arrived to find the fire controlled by the sprinkler system. Extinguishment completed by the fire fighters using hose.
Merseyside	Fire in an air conditioner through bearing seizure.	Sprinkler extinguished the fire. Air conditioner repaired and operating within two hours.
East Anglia	Workshop fire.	Fire controlled by a sprinkler. Damage minor.

Sheffield	Restaurant fire.	Sprinkler system controlled the fire. Restaurant re-opened after two hours.
London	Hotel fire, caused by rubbish ignited by a cigarette.	One **sprinkler head** came into operation and contained the fire.
West Yorkshire	Fire during textile manufacturing, overheated materials caused nearby cardboard packaging to ignite.	The **sprinkler heads** came into operation and controlled the fire. Resumption of full operations within a very short time.
Northamptonshire	Fire in a supermarket.	Sprinkler system activated, hardly any work for the fire brigade to do when it arrived!
Merseyside	Fire involving clothing racks.	Sprinkler system containing the fire by the time of arrival of the fire brigade, who completed the extinguishment without difficulty.
Northamptonshire	Fire in the entrance area of a block of flats.	Two **sprinkler heads** came into operation. Only loss a "wheelie bin". Those evacuated able to re-enter after about an hour.
West Yorkshire	Fire involving varnish and paint.	Eighteen **sprinkler heads** activated.

Sprinkler versus fire hose: relative amounts of water for a particular extinguishment.

It has been mentioned in a number of entries, including that for the **Chicago Hilton**, that it sometimes happens that sprinklers extinguish totally the fire at which they are directed by the time the fire department arrive. There is an advantage to this additional to that of rapidity: fire hoses when applied to small fires release excessive water. In general, other things being equal, a sprinkler will use about 1% of the water that a fire hose connected to a **vehicle-mounted pump** would. The result is water damage when the hose is applied.

Stadis® 450

Fuel conductivity improver. A DuPont product, it is added to aviation turbine fuel (ATF) at up to about 600 ppm.

Standpipe

Device for taking water from a below-ground water hydrant for attachment to hose. In the UK a standpipe will be of height 30 to 33 inches and, consistently with widespread use of **London Round Thread** outlets, will have a 2.5 inch connector for attachment to the hydrant. There might be a single head for hose attachment or a double one. Aluminium construction is fairly common and certain alloys are also used. In the US where fire hydrants at street level are above ground, standpipes are used for fire fighting inside multi-storey buildings where they might be connected to the building's own water supply ("wet standpipe") or need water supply from the fire department ("dry standpipe"). They are likely to be installed rather than attached for use in the event of a fire, as with a "standpipe" in the UK sense. It is believed that there was standpipe failure at the recent fire at the **Deutsche Bank Tower, New York**.

Stanford

Tug constructed in Holland, owned by BP with which company it entered service in 2005. It is equipped with fire fighting equipment to **FiFi I** level as is its sister ship *Castle Hill*. The vessels have been in service close to the **Coryton refinery** for duties including tanker handling. Approximately contemporary with these vessels are **Smit Kamara** and her two sister ships.

Staten Island, explosion at in 2003

In that year, when 9/11 was recent and Staten Island was on "high alert", there was an explosion involving a vessel taking on a payload of gasoline at a terminal there operated by ExxonMobil. There were two fatalities and one non-fatal injury. The vessel, which sank after the explosion, was holding 100 000 barrels ($\approx$4 million US gallons) at ignition.

Station Nightclub

Venue in Rhode Island where in 2003 there was a fire which claimed 100 lives. There were non-fatal injuries to 200 persons. The following points apropos of the fire can be noted. The club did not have an automatic sprinkler system. It was not exceeding its legal maximum number of occupants at the time of the fire. Although there were the legally required number of exits, persons inside the building disregarded these and tried to use the main entrance as an escape route instead; this is believed to have been a major factor in the high number of deaths. In terms of loss of life it was the second worst nightclub fire in US history: the worst was that at **Cocoanut Grove**.

Starbeck, Yorkshire

The scene of a recent bakery fire. An electrical fault is believed to have been the origin as at the Bakery fire in **Elkland, PA**. However, natural gas leakage was observed during the fire fighting and this required engineers from the gas company to attend to turn off the supply in order to protect fire fighters from a gas explosion.

Starship Majestic

This vessel, then owned by Disney World, experienced a fire during a voyage from the Bahamas to Florida in 1991. The fire began in the engine room. The vessel had originally entered service in 1972 as the P&O ship *Spirit of London*, having evacuation lifeboats closer to the waterline than was usual at that time. The ship was repaired after the 1991 fire and is at the time of writing in service as the *New Flamenco*.

State Fire Marshal, fusion of with other posts

Such fusion applies in a number of states of the US. For example, in North Carolina the posts of Commissioner of Insurance and State Fire Marshal are combined and the holder of the dual post is elected by the voters of the state along with other members of the North Carolina Council of State. The present incumbent of the dual post, James E. Long, has been in office for 24 years having therefore been re-elected several times. In Florida the posts of State Fire Marshal and Chief Financial Officer are combined and again appointment is by election. The present office holder is Alex Sink, who brought an already considerable profile to the post. In Mississippi the Commissioner of Insurance and the State Fire Marshal are one and the same individual.

Steam, static hazards from

When saturated (as opposed to superheated) steam has a dryness fraction significantly below unity, static electricity can be generated by contact between liquid droplets. This can be an ignition hazard if steam is used to purge tanks having contained hydrocarbons or if steam leaks into such tanks. This is believed to have caused the explosion in the *Fiona*.

Stedair®

Protective fabric for fire fighter apparel having a three-layer structure, the middle layer being a **moisture barrier**. The **THL** of the material is in excess of 650 Wm^{-2}, well in excess of the NFPA minimum of 450 Wm^{-2}. The manufacturer is the Canadian company Stedfast.

Steel for building construction, protection of with intumescent coatings

Intumescent materials including **Chartek 7** have found wide application to steel supports in buildings, where their action is in extending the period over which the steel will fulfil its structural role by slowing down heat transfer from a surrounding fire. One of the many recommendations to have been made as a result of 9/11 is that steel for use in buildings should be treated with an intumescent coating, not *in situ* once mounted in a building, but previously. The primary benefit is improved quality control when the retardant is applied at ground level, the result being easily examinable afterwards. There are spin-off advantages, including reduced demand on scaffolding and simplified scheduling of operations on site as one such operation is in effect taken off site.

Stenor vulcaniser

Device by means of which a jacketed fire hose can be repaired after mechanical damage. Its use requires skill and involves several steps. After thorough drying of the hose inside and outside, a hole penetrating the hose layer structure is made around the site of the damage. Through this hole a drill bit is inserted, which when pulled in a direction normal to the inside surface of the hose at the point of contact provides an abrasive surface. The bit in this position is rotated at about 1000 rpm and this prepares the inner surface for the repair. (The procedure is called "hose scouring".) The damaged part of the outer layer is prepared by a wire brush attachment to the drill. A patch for repair of the internal surface is manoeuvred into position and the vulcaniser is brought into operation at a temperature of 165 °C. A second patch is then placed on the outer surface of the hose in a position corresponding to that of the patch on the inner layer, and the vulcanising is repeated. The overall effect is consolidation of the inner liner at the position of the damage and the inner and outer patches into a single repair plug. A damaged hose can in this way be returned to its original operating pressure specification. There are alternatives to the Stenor product on the market including the Pronto 10 hose vulcaniser. Where hose damage is to the inner layer only, repairs using a single patch and adhesive instead of vulcanising are sometimes possible.

Sterilising of cloths in a microwave, dangers of

A few years ago advice, originating in the US, that to place a cloth in a microwave on full power would very effectively sterilise it was widely circulated. This is in fact dangerous practice. It is well known that a microwave operates by absorption of the microwaves by water. If there is no water, the microwaves will undergo multiple reflections leading to arcing, an extreme fire hazard. That is of course why a microwave must not be operated when empty. If a cloth placed inside a microwave has only a moisture content in equilibrium with the natural humidity of the room in which it was previously standing, there is a risk that there will be insufficient moisture and an arcing effect. In early 2007 Cleveland Fire Brigade in the UK responded to a fire which had been caused in that way.

Stilan

Trade name of a polyarylene insulating material. It was used for electrical wires in Boeing 747s over the first five to ten years of their manufacture, also in DC-10s.

Stopseal®

Fire collar, suitable for pipe diameters spanning almost an order of magnitude in diameter with a fire rating of up to four hours. The intumescent material is graphite based and is held in a stainless steel cylindrical support. Stopseal® collars find application to **uPVC** pipes and also to steel pipes the contents of which are flammable.

Storex Laylite

Manufactured by Canon, an example of fire hose with a **textile-reinforced rubber** structure. Developed for use with fire trucks, it comes with blades in the diameter range 45 to 64 mm. The precise value depending on the blade size (the blade rotation rate is fixed), air movement is of the order of tonnes per hour.

Storz coupling

Device by means of which the two parts of a connector are equivalent and identical (as, of course, they are not with a screw thread) and can be linked together by a quarter turn very quickly. It has found very wide application in fire fighting, in the connection of hose in hydrants and in the **water thief**.

Striker

ARFF vehicle built by the Oshkosh Truck Corporation in Wisconsin (which is in the same group of companies as Pierce) and used by the USAF as well as at civilian airports. Boeing also use it at their manufacturing base. All versions of the Striker are obtainable with a **Snozzle™**. Available with two, three or four axles, a Striker will carry water, foam, dry chemical agents and **Halotron™**. The power unit is in the range 650 to 950 hp. The high power provides acceleration for rapid response and also the ability to surmount obstacles in order to reach a crashed plane.

Stril Poseidon

Anchor handling, tug and supply vessel having **FiFi I/II** classification. Registered to Norway, it operates in the Haltenbank fields, which are being developed by Statoil. The *Viking Poseidon*, also registered to Norway, is a supply vessel only and as such does not have fire fighting capabilities. This vessel is in fact scheduled to undergo a conversion to a seismic research facility.

Styrene–butadiene rubber (SBR)

Material suitable for the fabrication of hose for interior use, sometimes having a layered structure and textile reinforcement as with **Sirocco 9303**, which uses not SBR but thermoplastic elastomer (TPE). A hose made from SBR of the configuration described operates at moderate pressures, up to about 220 psi.

Subsidiarity in fire legislation

A law or ordinance to which subsidiarity applies is to be implemented by a lower level of authority than the that of legislature which brought the law or ordinance into being in the first place. The question of how any law which the European Parliament migh pass on sprinklering of hotels would be enforceable, if at all, in member countries was raised after the fire at the **Penhallow Hotel**. In the US it has been asserted (not necessarily correctly: results of test cases are awaited) that Federal legislation relating to fire sprinklers is relegated to the Executive and Judicial functions of the respective State governments. An interesting aside is that since 1990 the US Government has required that Federal employees travelling on business and requiring an overnight stay in any state must use a hotel registered as being "sprinklered".

Sumitomo Chemical Company

Destruction in 1993 of this company's factory at Niihama on Shikoku Island caused a worldwide disruption to supplies of epoxy resin. Especially affected were integrated circuit manufacturers. There was one fatality from the fire.

Summit tunnel fire

A few days before Christmas in 1984, a goods train bearing tankers of gasoline was passing through a tunnel close to the north of England town of Todmorden when one of the tankers derailed. There was a domino effect whereby the tankers behind the one having derailed did likewise. There was almost immediate ignition of the leaked gasoline. At the direction of the attending fire services, the railway crew removed the remaining tankers from the tunnel, drawing them out with the locomotive, and fire fighting equipment was sent into the tunnel. Hoses were lowered into the tunnel through the ventilation shafts. Fire fighting continued until the crews perceived, correctly, an incipient explosion in one of the overturned tankers and evacuated just in time. In the escalation which followed, flames were to be seen above the (quite deep) ventilation shafts and there was spread of fire to vegetation necessitating closure of a major road. Fire fighting continued for four days and the fire services maintained a presence at the site for a full week into January in case fire redeveloped. Remarkably, damage was not severe, consisting mainly of destruction along about half a mile of track and electrical installations. In terms of the total quantity of heat released over its duration, the Summit tunnel fire is probably the worst transport tunnel fire ever.

Sun Venus

Tanker carrying ethanol, upon which there was a fatal explosion close to the Japanese coast with loss of life. Static electricity has been suggested as a cause, as it has in the case of the US Tank Barge *STC 410*, which whilst jet fuel was being taken from it exploded killing four people. A PVC device ("wand") had been used in the fuel offloading and it is thought that this might have become charged.

Super Tiller

Tractor-drawn aerial from E-One with a ladder reach of 100 feet. Developed for Kansas City, who placed an initial order for five, it is now used by other US fire departments including that at Baltimore County, MD. It is powered by a Detroit Diesel unit.

SuperVac Negative

Fan developed for negitive pressure ventilation (**NPV**), like many such using electricity instead of a gasoline or diesel engine as is usual with fans for positive pressure ventilation (**PPV**). Models are available with fan blade diameters in the range 12 to 24 inches, giving air flow rates of typically 12 000 m^3 per hour of air which in weight terms is about 13 tonnes per hour.

Supply line holder

Simple device whereby the hose from a **water tender** is attached to a **dump tank** during transfer of water from one to the other, eliminating the need for it to be hand held and so freeing up a member of the fire fighting team. Commonly made of PVC and of 2 to 3 inch exit diameter, selling for about US$200.

Sutphen

Manufacturer of fire appliances, based in Ohio. It is perhaps specially noted for its range of **quints**. An example is the Sutphen SP110, which carries 500 gallons of water and has aluminium **ground ladders**. Its aerial ladder has a reach of 110 feet and it carries 1000 feet of hose. It is usually supplied with a pump but sometimes not at the specification of the purchaser in order to make for more compartment space. So configured, it does not conform to the *precise* definition of a quint.[36] An SP110 having a pump was delivered to the Fire Department in Galena, OH in May 2007. Others in the Sutphen range include the SP90, with an aerial ladder reach of 90 feet. All in the Sutphen range of quints have stainless steel body panels, which makes for longevity. Sutphen run a large upgrading and retrofitting facility, whereby an appliance dated in terms of it equipment but structurally sound can be brought up to current standards at a cost much lower than that of a whole new appliance.

Sweep purging

Continuous flow of inert gas through the space occupied by the vapour at only just above atmospheric pressure until the target oxygen concentration is reached. A depleted storage vessel of liquid is likely still to contain vapour, possibly with air in proportions such that ignition is possible. Such a vapour/air mixture can be removed by admitting to the vessel as much water as it will take, whereupon the vapour/air will be confined to the small space above the surface of the water and can be removed with a single "sweep purge". As the water is drained out it can be replaced by inert gas, leaving an inerted empty vessel ready for service.

Syltherm

Heat transfer fluid, containing both carbon and silicon (a dimethyl polysiloxane). It has a flash point of 77°C. There are, however, incidences of reduction in the flash point of Syltherm during use and these have been attributed to skeletal rearrangement of the structure during use at higher temperatures. This effect has been observed by users and is noted on the Materials Safety Data Sheet.

---

36 The term is sometimes used by fire departments to mean an aerial ladder/pump combination.

Table Mountain fire, 2006

This occurred in late January that year, and its propagation was aided by strong winds and very dry conditions. It spread to suburbs which occupy the northern ascent to the mountain. There was one fatality, a female British tourist who died from smoke inhalation. Loss of vegetation, including certain species almost unique to the area, was heavy. Helicopters as well as ground fire crews were involved in the fire fighting. An electricity mains cable was destroyed, leaving some residential areas without electricity. A British national was arrested on a charge of having started the fire by discarding a cigarette. He was subsequently acquitted.

Taegu

In February 2003 this Korean city was the scene of a subway fire which claimed 192 lives. An individual with a history of mental illness set fire to the interior of one of the six carriages comprising a train, using gasoline as an accelerant. About four litres of gasoline were so used. Only two minutes after ignition within one of the carriages the other five were on fire, and there was spread to another train also with six carriages, so twelve carriages in all were affected. There were fatalities in both trains. Spread of the fire from one train to another was made possible by the fact that the space between them was occupied by flammable materials. These included seats, flooring and advertisement boards made of combustible materials, including polyethylene.

Takarazuka

Locality in the Hyogo Prefecture of Japan, the scene in 2007 of a fire at a karaoke bar which killed six teenagers. An employee at the karaoke bar was subsequently arrested for negligence, having left an open pan containing cooking oil unattended. This is believed to have been the origin of the fire. Even more grave in its effects was the karaoke fire at **Inchon**.

Talisman Wasp

Thermal imaging camera for use in fire fighting. It runs on disposable batteries and can be interfaced with a **SCBA**.

Tanker 910

Term applied to a particular **air tanker,** based on the McDonnell Douglas DC-10 and the only one of its kind currently in service. It was built by adapting a DC-10 previously in service as a passenger airliner. It can carry up to 45 600 litres of water, making it equivalent in those terms to the **Ilyushin IL-76**. At the present time Tanker 910 is used solely by CAL FIRE for forest fire control in California. CAL FIRE pay the owner of the aircraft (2007 rates) just over US$5 million for its exclusive availability for 122 days in the year and US$5500 per hour's operation in the air.

Tan oak

A.k.a. tanbark oak, botanical name *Lithocarpus densiflorus*. It occurs in California and in southern Oregon and is classified as a **highly flammable tree**.

"Teakettle"

Name by which the first fire truck in Piqua, OH was popularly known, although its correct name was Reliance 1. Having been sourced in Columbus at a cost of US$1075 it entered service in Piqua IN 1839. It had 250 feet of hose and was retrofitted with ladders in 1845. In that same year three "fire wells" were dug in Piqua to provide a water supply for fire extinguishing. Each fire well had a manual pump to raise water. Piqua, current population *circa* 21000, now has a fire department with four fire trucks of combined water application capacity ≈ 1500 litres per minute.

Tefzel®

Dupont product, a modified form of ETFE (ethylene–tetrafluoroethylene). Its uses include electrical wiring insulation in aircraft. It is soft and gets more so on heating, so its use in a bundle of wires is deprecated.

Tej shoe factory, fire at

This Indian footwear factory operated by Shree Jee International was the scene of a major fire in May 2002. The death toll was 43 and there were several injuries. There were only two exits at ground level, one of which was locked at the time of the fire. The fire was due to ignition of chemicals used at the factory. Whether the ignition source was an electrically created spark or a discarded match or cigarette is not known. A follow-up revealed negligence in criminal degree on the part of management. The factory was in an export processing zone (EPZ): these exist in many countries other than India, including China and Cuba. At an EPZ foreign companies operate their own manufacturing facilities using local labour and there is waiver of export tariffs. There has been extreme concern, articulated for example by international trades union organisations, about health and safety conditions in EPZs.

Tellus 32

Hydraulic oil manufactured by Shell Canada for use in low-temperature conditions. The product data sheet states that its flash point is above 100°C. Applications include mining machinery.

Tetrabromobisphenol A (TBBPA)

A brominated fire retardant, molecular formula $C_{15}H_{12}Br_4O_2$. Its primary uses are in printed circuit boards and television sets. It is incorporated into such at levels of about 0.05%. Annual production of TBBPA worldwide is 120 000 tonnes. In the US, domestic production is supplemented by import from Israel. Canada and Germany also make TBBPA for export. There is major production in China by "western" companies who have established a presence there. Both WHO and the EU have evaluated TBBPA for health effects when used for the purposes outlined above and have found no reason for concern.

Tetrakis(hydroxymethyl)phosphonium hydroxide (THPOH)

This reagent when reacted with ammonia forms a flame retardant for cotton fabric. Cotton is first soaked in a solution of THPOH and treatment with ammonia follows. There are many fire-retardant cotton fabrics having registered trade names which have been manufactured in this way, including Amtex® and **Excel-FR™**.

Tetramethylolphosphonium chloride (THPC)

Substance used (like **THPOH**) to render cotton fabric fire resistant. Being the salt of a strong acid and a weak base, THPC is quite acidic, making pH control important in its use. In the fire protection of cotton it is applied with urea, and the final chemical step is oxidation of phosphorus from +3 to +5 with hydrogen peroxide as the oxidising agent. A variant on this procedure is the **Proban®** process, in which there is also treatment with ammonia. The sulphate salt (THPS) can be used instead of the chloride. Also, in application to fabric an amide is sometimes used instead of urea.

Texaco Rando® HDZ

Series of six hydraulic oils, denoted by numbers, with flash points in parentthesies: 15 (150°C), 22 (188°C), 32 (220°C), 46 (186°C), 68 (212°C) and 100 (232°C). As the series number increases, so does the kinematic viscosity, the most important single quantity in the performance of a hydraulic oil and the basis of choice for a particular application. Across the range of viscosities the flash point fluctuates, though all have high flash points. Oils in the **Nuto H** range do have flash points which range, though certainly not widely, being in about the same range as those for the Texaco Rando® HDZ series.

"Texas Snow Job"

A **compressed air foam** extinguishing agent which, having been successfully developed in 1975, predates the recent trend towards **CAF** systems by 20 to 30 years. Developed by the Texas Forest Service, it used as foam concentrate a pine derivative obtained from the local paper industry. It used air from cylinders, and this imposed a limit on operating time.

Textile flock, hazards of

The formal definition in the UK of textile flock is that it consists of fibres not exceeding 5 mm in length. An explosion ensued when flock was drawn into a gas-fired oven at Clarkson Textiles, Nelson, UK in January 2002, and one worker received serious burns. The flock had been drawn into the oven by the forced ventilation system and had ignited in the oven's pilot flame. The Company were fined £9000.

Textile-reinforced rubber

This provides an alternative to the jacketed structure of most fire hose which, as described in other entries, has an inner layer of rubber or equivalent synthetic material and one or two surrounding woven layers. The textile-reinforced rubber uses rubber into which a textile material has been incorporated, and on examination of a cross section of such a hose a single layer is evident. An obvious advantage is that laying flat for storage is possible for wider diameters than with jacketed hose.

TGS 6812

Catalytic gas sensor from Figaro in Illinois, intended for detection of hydrogen and of light hydrocarbons. It incorporates a feature whereby polar organic molecules such as ethanol are adsorbed before they can contact the catalysts pellet and therefore cannot interfere with the readings. Such interference is known as "cross sensitivity".

TGS 813

Combustible gas detector manufactured by Figaro. Using a **tin(IV) oxide** semiconductor, it responds to carbon monoxide, hydrogen and alkanes up to C_4. Other things being equal, the carbon monoxide response is the strongest. The semiconductor is held at the required temperature by means of a 5 V supply which can be AC or DC.

Thatched roofs, fire precautions with

Three preliminary points are as follows. First, at least in the UK there is no statistical evidence that a thatched property is more likely to catch fire than one with a conventional roof. The difference is that propagation of burning once established is more rapid with a thatched roof. Secondly, some thatched properties are "listed", in which case substitution of a conventional roof for a thatched one is not possible. Such substitution is not necessarily encouraged in any case where thatched buildings conserve the character of a location. Thirdly, some fire brigades in areas where there are thatched buildings follow predetermined action on being informed of a fire in a thatched property. This means that more appliances and fire fighters are sent than would be to a regular "house fire". By far the majority of fires in thatched properties are due to chimneys. Guidelines therefore apply as to how high the exit from a chimney must be above the thatched roof: five feet is seen as a minimum. Close attention must be paid to the condition of the chimney where it passes through the thatch, and insulation applied. Electrical wiring close to the inside roof needs more frequent safety checks than in the case of a conventional roof. When plumbing is taking place anywhere near the inside surface of the thatch, compression joints should be used instead of soldering or brazing. (See *also* **Maria Ratsehitz Mission Hospice.**)

ThermaCAM E2

Thermal imaging camera having found marine application. It is used to identify parts of a ship's interior surface, which for whatever reason have become heated. Such a precaution against fire is required under the Safety of Life at Sea (SOLAS) regulations.

Therm-A-Hinge

Wood, depending on the tree from which it came and on whether the measurement is made across the grain or along it, has a thermal conductivity of the order of 0.1 $Wm^{-1}K^{-1}$. Any metal from which hinges on a door are made will have a much higher value than that; for example, galvanised steel has a thermal conductivity of about 40 $Wm^{-1}K^{-1}$, about two-and-a-half orders of magnitude higher than that of the wood. This means that in a fire the hinges will provide for heat transfer to the wood, thermal breakdown products of which will add fuel to the fire. Therm-A-Hinge is a device whereby a pad of a material displaying **intumescence** is placed between the hinge and the wood. Its swelling in a fire will coat the metallic hinge and inhibit heat transfer to it, therefore protecting the door. There are intumescent devices which perform the same function on door locks.

Thermal imaging

Means of detecting and determining the position of something on the basis of infrared emission from it. The detector, part of a thermal imaging camera, works on the principle of changes in the properties of a germanium "lens" in response to infrared absorption. It is used in fire fighting to locate the seat of the fire, that is where it began, and also areas of the fire of particularly high temperature and areas of residual combustion ("hot spots") after apparent extinguishment, as at **Leicester Paper Converters**. It can also be used to find persons where to do so visually is precluded by smoke. (*See also* **Franklin, IN**; **K1000 Elite**; **Talisman Wasp**; **ThermaCAM E2**.)

Therminol 66

Heat transfer fluid, recommended for use at up to 345°C. It is very resistant to chemical change even when operating over long periods at temperatures at the high end of its operating range. The flash point is 184°C.

Thermistor, use of in heat detectors

A thermistor is a device by means of which temperature can be measured on the basis of electrical resistance. A heat detector for fire protection purposes might consist of two identical thermistors. Any difference between the resistances of the two signifies heating and this is the basis of the signal and response. An alternative is for such a detector to use thermocouples. In a thermocouple circuit the resistance is of course irrelevant as the emf developed is independent of it, provided that the measuring device to which the emf is sent is of high impedance.

Thermo-Gel®

Gel concentrate. It is used in fire management in two ways. First it can be applied to buildings and structures when threatened by fire, for example a brush fire encroaching upon a residential area. It has been so applied to walls, floors, shrubs, cars and fuel tanks. Secondly, it can be used in fire fighting when it is added to the extinguishing water in Class A fires and in forest fires. In the latter the water containing Thermo-Gel can be dropped from the air, as recently happened in the Diablo mountain range in California. Thermo-Gel® contains polyacrylamide and sodium polyacrylate. A functionally similar product is **AFG Firewall™**.

Thistle Alpha

North Sea oil production platform on which in November 2007 there was a fire which necessitated evacuation of the majority of workers on board. There were no deaths or injuries. The rig is operated by Petrofac.

THL

Total heat loss, from a fire fighter's body whilst on duty at a fire. The higher the THL the better. In materials and devices for fire fighters a good THL has to be accompanied by a good thermal protection and some factors promoting one can jeopardise the other. Both are addressed in the specifications of such materials as **Crosstech®** and **Stedair®**. The current minimum THL for such materials set by NFPA is 450 Wm^{-2}.

ThunderFog 250

Adjustable gallonage nozzle which operates at a fixed pressure of 100 psi and will deliver water in fog form at rates of 95, 125, 150, 200 and 250 gpm according to the flow disc selected. This nozzle is frequently used with **AFFF**.

Tidalwave 600™

Double jacket fire hose, like **Rack Hose™** having a lining of thermoplastic polyurethane (TPU). The jacket is made of polyester. Tidalwave 600™ is suitable for use in attack or for supplying an attack hose. It comes in 6 inch diameter.

Tiger ship fire, 1613

Manhattan is surely one of the most prestigious and glamorous parts of the world and one of the most expensive in terms of real estate. That it was ever selected for European habitation is in fact a spin-off from an accidental fire. In 1613 the *Tiger*, a Dutch vessel captained by Adrian Block, sailed into waters close to the present New York. Block had trained in law in Holland and was an enterprising individual; he planned that the *Tiger* would return to Europe laden with furs for sale there. Having collected some cargo by way of beaver and otter pelts, he berthed his vessel at the southern tip of what is now Manhattan Island. (Four years previously another Dutch explorer had in his journal given it the name Manahata.) Block and his crew disembarked and set up a camp on the Island, at which stage the *Tiger* caught fire and was destroyed. The consequences were of course much more serious than simply loss of vessel: Block and his men were stranded thousands of miles from home in an environment which was hostile climatically and in other ways. Over the next few months they built another ship, with wood felled locally and using a few tools salvaged from the *Tiger*. Whilst timber was plentiful, the nails to hold it together were few, only those salvageable from the *Tiger*. (It is not known from what material the sails were made.) The newly built vessel was named *Restless*. Being the first Europeans to have resided in the area, Block and his men established trade with the indigenous people. Wampum (decorative shells) obtained by the Europeans were exchanged for furs provided by the Mohawks. This trade established a commercial base for subsequent European arrivals and so Manahata/Manhattan entered development.

Tin(IV) oxide

The tin oxide semiconductor finds application to gas sensing, for example carbon monoxide. For this purpose it is on a platinum base and heated. Contact of gas molecules alters the conductivity of the oxide and this is the basis of the signal. Tin oxide detectors are also used for methane. A semiconductor can be adapted to particular gases according to the nature and quantity of the dopant and the temperature. A tin oxide detector needs to stabilise at its operating temperature for a time of the order of days before being put into service.

Titan

Range of **ARFF** vehicles from E-One. The 4 × 4 version uses a power unit manufactured by Detroit Diesel which develops 665 hp and has a pump supplied by Hale capable of delivering 1500 gpm of water. The water tank holds 1500 gallons and foam is also carried.

Titanic, coal bunkers on

Professor R.H. Essenhigh of Ohio State University has an accomplished background in combustion research and his theory on the sinking of the *Titanic* deserves a hearing. He sets out two established facts. First, that the vessel was sailing at full speed in an area known to contain icebergs. Secondly, the vessel could not have operated "full steam" for the entire journey since, because of a coal strike, she had departed with insufficient fuel to sustain full steam conditions for the entire crossing. In any case the appeal of the *Titanic* was its luxury rather than its speed and the latter would not have been a preoccupation of crew or passengers.

Essenhigh's view is that in one of the coal storage bunkers the coal was burning, this of course threatening the ship if no action was taken. The combustion would have been well within the depth of the bunker, not at the surface, and before it could be reached coal distant from the burning zone had to be removed and admitted to the steam generators. Once the burning coal was accessible extinguishment with water was possible. So, Essenhigh argues, transfer of coal from bunker to steam generator as a fire mitigation measure had the indirect effect of accelerating the vessel and exacerbating the impact with an iceberg. It would of course have been possible simply to release the extra steam so generated so that it did not cause the vessel to speed up, but once coal spontaneous heating has begun, extinguishment measures only buy time, so the problem could not be reliably totally eliminated until the vessel had reached *terra firma*. That would have been a good reason to operate at high speed in the circumstances.

Titanium chloride

$TiCl_4$, flame-retardant substance for wool and applied to clothing, carpets and upholstery. Its application to wool requires a carboxylic acid, and citric acid is a common choice. It has the disadvantage of causing yellowing of the wool.

TKT

Composite material – **Kapton®** with an inner and an outer layer of Teflon – for aircraft wiring. It replaced Kapton® alone in Boeing aircraft. Airbus use KT, Kapton® with one layer of Teflon.

Toa Oil

This company, which came into being in 1924, is concerned with downstream operations and receives crude from Showa (Shell) amongst other oil companies. In 2006 there was an explosion of heavy petroleum material in a stationary tank at Kawasaki belonging to Toa Oil. The tank contained about 25 barrels of the material, which was being kept at 160°C to prevent sludge deposition.

Tobago, status of the fire fighting services in

This small island in the Caribbean, part of the nation of Trinidad and Tobago though having its own House of Assembly, is currently undergoing an urgent review of its fire service. In the capital Scarborough several hydrants were rendered inoperative by recent pavement reconstruction and in other parts of the island some hydrants have on inspection been found to be leaking water. Many residential buildings in Tobago are crudely constructed without regard to fire codes. The Tobago fire service does not have its full complement of officers. Trinidad and Tobago is experiencing prosperity at the present time through its offshore gas fields. Two major overseas companies – BP and Methanex – are investing heavily and Tobago, insofar as it can do anything independently, is pressing for its share in the revenue from the gas. The raising of standards of municipal services will, one hopes, be the result.

Tokyo, night life of

Even high-income residents of Tokyo often live in small apartments and do not want to spend all of their leisure hours within them. Consequently, participants in Tokyo's night life include a high proportion of residents in addition to visitors and such residents are imperilled if fire protection at nightspots is substandard. Recently concern has been expressed in relation to karaoke venues in particular, partly as a result of the fire at **Takarazuka**.

Tomakomai

Japan has a refining capacity of the order of 5 million barrels per day, having built a large number of refineries during the reconstruction period after the Second World War. A refinery at Tomakomai on Hakkaido Island, belonging to Idemitsu Kosan, was the scene in 2003 of a fire resulting from an earthquake. Naphtha being stored in a tank spilt and ignited and three local residents were treated in hospital for smoke inhalation.

Total area discharge method

In the use of carbon dioxide for fire extinguishment, the technique of sealing off an area enclosable by doors and admitting carbon dioxide to it to protect from fire at the other side of one or more of the doors. The method can also be used where an actual fire and an area which could be affected by it are separated vertically, that is by a floor.

Total release insect sprays

Chlorofluorocarbons were once used as propellant for insect sprays but have been phased out because of their effect on the ozone layer. They have been replaced by hydrocarbons such as propane, and in a "total release" application this can result in enough hydrocarbon for the lower explosibility limit to be exceeded. At least one house in the US has been totally destroyed in this way.

ToughStore

Range of cabinets for storing extinguishers and hose at offshore platforms, from the same manufacturer as the **Firebird** range and even more robust than the latter. Whereas Firebird products are made from **GRP**, ToughStore products contain ABS (acrylonitrile–butadiene–styrene) and have transparent doors made of polycarbonate.

Toy-like lighters

Lighters having the appearance of a toy, for example in the image of a dog with the flame issuing from its "mouth", are available is some parts of the US. The **NASFM** have petitioned the Consumer Product Safety Corporation to have them prohibited.

Toxteth

Inner-city district of Liverpool, England where in August 2007 there was a fatal vehicle fire. A van being pursued by a police car exploded and the van driver died. The probable explanation is that fire began through fuel leakage or by electrical malfunction, exploding when the contents of the fuel tank were ignited.

TPO

Thermoplastic polyolefin, based on polypropylene and ethylene–propylene rubber (EPR). It finds wide application to roofing, when it is treated with a suitable colouring agent and also a fire retardant (often magnesium oxide). In the roofing material trade one of its competitors is **Hypalon®**.

Tractor-drawn aerial (TDA)

A.k.a. a tractor-drawn platform, an appliance with aerial ladders having an articulated arrangement whereby the cab can move, within a certain angle range, relative to the rear of the truck during driving. This of course makes for enhanced manoeuvrability in heavy traffic and narrow streets. **Seagrave** manufacture TDAs and one of their most recent, having a ladder reach of 100 feet, was delivered to the Fire Department in Somerville, MA in October 2007. American LaFrance also manufacture TDAs and have supplied customers including the San Francisco Fire Department. Pierce, whose products feature in a number of other entries in this volume, including that for the **Contender**, have recently received a contract to build a TDA for Raleigh, NC. **Crimson** are also a manufacturer, having recently delivered a TDA to the Fire Department in Anaheim, CA.

Transformer oil

Having a cooling and an insulating role, transformer oil is usually closely specified distillate from crude oil although some, such as **Midel 7131**, contain synthetic esters. A distillate transformer oil will however be from the higher boiling end of the range and will have a flash point comparable to that of some diesels and well above room temperature. Nevertheless, transformer oil fires are very common and their frequency in the US is approximately one per day. Use of polychlorobiphenyls (PCBs) in transformers has declined because of their harmful environmental effects. These are of course ignitable, and the ease of ignition decreases with increasing degree of chlorination.

Trawler fire, 2007

There was loss of life when the Faroese trawler *Hercules* experienced a fire originating in the engine room whilst working in Chilean waters in April 2007. The survivors were taken to safety by other vessels in the area. (*See also* **Gerda Maira**.)

TREMstop WS

Graphite-containing intumescent material from Tremco, manufacturers of **Caulk®** and **Walk**. It is supplied in strips for application to the surface it is protecting, and expands by a factor of 10 when heated. TREMstop WS is one of a family of products which includes TREMstop Strap, which acts similarly to a **fire collar**. It is installed along a circumference of a non-metallic pipe, which it protects by **intumescence** when the pipe becomes heated. (*See also* **Fire pillow**.)

Triangle shirtwaist factory fire

Occurring on 25 March 1911, a fire at this clothing factory in New York City claimed 146 lives. At the time of the fire there were about 500 employees, mainly migrant women. Typically for that period, housekeeping was poor and bits of combustible textile materials littered the floors. Illumination was by gas and smoking was prevalent amongst the workers. After the fire had begun, it was discovered that one of the doors which ought to have provided an exit route was locked. There was a fire escape made of iron, but it was anchored to the building in a most inadequate way and soon became detached once persons attempting to escape the fire started to use it.

The owners of the factory were inside it when the fire began, and both escaped injury. They were put on trial and acquitted, although they had to pay compensation as a result of a subsequent civil action. The factory was not destroyed in the fire, and the building is in fact still there having since 1929 been part of New York University.

Tribune Building

This building in Chicago, completed in 1925, is soon to be retrofitted with sprinklers. This is part of a renovation program for the building. The local ordinance requires that in high-rise buildings built prior to 1975 installation of sprinkler systems begin in 2005 and be complete by 2017. The Tribune Building will therefore, once the sprinklers are in place, have exceeded this requirement by a considerable margin. Installation of the sprinklers over a short period instead of over the 12-year period stated in the local ordinance has been made possible by working in the installation with a general overhaul of the building. In fact, the status of the Tribune Building as a "designated landmark" might have made it eligible for exemption from the ordinance, subject to approval of some alternative proposal.

Tricresyl phosphate

Molecular formula $C_{21}H_{21}O_4P$, used as a plasticising agent, for example in PVC, also having a role as a fire retardant. It also finds application to polyethylene and to resins, adhesives and rubber. A major manufacturer of tricresyl phosphate is the Jiangsu Cahngyu Chemical Co., China, which exports the substance to Europe and America as well as supplying the home market. Fire Safety Standards in China are undergoing badly needed improvement and a factor in the progress of such improvement is lack of locally produced alumina trihydrate based fire retardants. This is being offset largely by use of brominated retardants but phosphate retardants also have a role to play. Further evidence of this is the manufacture of **Fyrol™** products in Shanghai. Another trade name under which tricresyl phosphate is marketed is **Disflamoll®** TKP.

Tridol S 3LT

Concentrate for making **AFFF**, having a fluorine-containing surfactant. It is used at airports in "rapid intervention" vehicles and also at fuel depots and at offshore platforms. It also finds application to extinguishment fires where in some circumstances it offers an alternative to an **ABC powder extinguisher**.

Trieste, refinery fire in 1972

There was sabotage of pipe work at a tank of crude oil at a refinery in Trieste in August 1972. The **internal floating roof** eventually failed (as happened at **Milford Haven**) and **boilover** resulted.

Triphenyl phosphate

Organophosphorus flame retardant like **tricresyl phosphate**, having a dual role as plasticiser and retardant, although in some applications its intended function is that of a retardant only, there being another plasticiser present. Triphenyl phosphate finds wide application to cellulose acetate products: other substances acceptable for use with cellulose acetate in terms of their performance as plasticisers, including phthalate plasticisers, tend to enhance its flammability. Triphenyl phosphate is used similarly with cellulose acetate butyrate. One trade name for it is **Disflamoll®** TP.

Trolley unit

Term applied to one or more extinguishers mounted on a trolley, "mobile extinguisher unit" being a synonym. Sometimes more than one type of extinguisher, e.g. a water extinguisher and a dry powder one, are carried on a single unit. Trolley units find wide application, construction sites being one of the major ones. Trolley units for such use often carry supplementary equipment, including a fire alarm and a fire bucket.

Turbine Thrush

Aircraft suitable for fire fighting having a water holding capacity similar to that of the **Dromadear**. It is manufactured in Georgia, USA and is a single engine aircraft. The Turbo Thrush showed its value in fire fighting in the forests of Natal in the early 1980s, having previously been in such use in Orange Free State.

Turbulence, effect of in fire hose

In general, the lower the degree of turbulence of water passing along a fire hose the better. This is because the higher the turbulence the greater the coefficient of friction. A **fixed gallonage nozzle** requires a set pressure in order to operate. It could happen because of turbulence that such a nozzle experiencing the required pressure would not, other things being equal, experience it on a longer hose because of the frictional effects. Considerable R&D has gone into reducing such effects in hose liner materials.

Turret

Strictly "turret monitor", aspirated or non-aspirated device for water application from an **ARFF** vehicle. An ARFF vehicle will have a bumper turret and a roof turret, these terms being self-explanatory. A turret can move in the same way that an **oscillating monitor** does. Akron, whose products include the **Vari-Nozzle**, offer a range of turrets for ARFF use. As pointed out in an earlier entry, there is much "ad hocery" in terminology for fire fighting equipment: the term "turret" sometimes means an ARFF vehicle itself. Such usage is evident on the websites of some manufacturers. (See also **Rhino II**.)

Twaron®

A fire hose, especially one of narrow diameter, can kink. In usage this makes for obvious difficulty and in storage it can lead to premature wear or breakage. In hoses made of materials such as PVC or **nitrile**, kink can be eliminated by incorporation longitudinally of Twaron® into the hose structure. Twaron® is one of the synthetic fibres generically referred to as para-aramid fibres. Some para-aramid fibres find application to fire-resistant clothing.

Topics in Environmental and Safety Aspects of Combustion Technology

Dr. J.C. Jones, Senior Lecturer, University of Aberdeen, UK

A selection of combustion topics that relate to safety and the environment all linked by combustion chemistry. The text contains many worked examples and case studies making it ideal for postgraduates and advanced undergraduates in chemical engineering, fuel and energy technology, safety engineering and related disciplines.

...the book can be highly recommended for interested professionals as well as for both undergraduate and postgraduate courses... *Fuel*

This book will be of particular value to postgraduates and final year students in chemical engineering, fuel and energy technology and associated subjects, including safety. ...it is highly recommended. ... Dr. Jones has presented matter by way of many worked examples and case studies which help considerably in understanding the underlying theme... *Energy World*

ISBN 978-1870325-66-0   234 × 156 mm   192pp   illustrated   softback   £17.95

U

Ubiko T2-4

Robotic device for sensing smoke, developed in Japan. It can assist in fire protection by patrolling buildings.

UK fire brigades, introduction of PPV by

Many UK fire brigades are still undecided about **PPV** and its possible introduction features in the agendas of management team meetings. One factor is the training time involved, even when PPV is limited to "post-incident use" (a.k.a. overhaul), where it is an alternative to **NPV**. Some brigades are training crew members initially in post incident use only and for use of PPV in attack later, the span of the training scheme being three years. Fire brigades which have adopted PPV, including Humberside and West Sussex, have reported using it to good effect in house fires. The fact that the West Sussex brigade is using PPV is an example of its adoption by a small brigade, whilst many large brigades are undecided. There are parallels to this in the US, for example **St George, UT**.

UK, fire hydrants in

In contrast to those in the US fire hydrants in the UK are positioned underground, access being via a cover which can be raised. A **standpipe** is required between the outlet from the hydrant and the fire fighting hose. Although all hydrants are required to conform to a standard (BS 750), there is scope for variation in some features, including the structure of the outlet for standpipe connection. Major cities have differed from each other in such details and place names have entered the terminology of the subject, for example the **London Round Thread (LRT) outlet** and the Dublin Bayonet outlet. There are several manufacturers and suppliers and any can bid for a sale provided that the product conforms to BS 750. Typically the hydrant structure will be cast iron with an epoxy-based anticorrosion layer. A pressure rating of 16 bar is common. The outlet might be made of a different material from the body e.g. gun metal or possibly plastic. There

will be a frost plug, by means of which water inside the hydrant after operation can be drained out making the hydrant analogous to the dry barrel
type of above-ground device. Whereas with above-ground hydrants such
as those of the **American Darling** series will have more than one outlet,
a below-ground outlet has a single one, although it can be attached to a
double-outlet standpipe.

Ulsan Refinery

The Ulsan Refinery in South Korea was the scene in November 2007 of a
fire which was readily brought under control. The refinery, with a capacity
of 840 000 barrels per day, is currently the second largest in the world.

Ultraguard

Alcohol-resistant aqueous film-forming foam (AR-AFFF) concentrate manufactured by Chemguard, whose HQ is in Mansfield, TX. Like foams from
other manufacturers (e.g. Kidde) having the AR prefix, it is suitable for polar
liquids although its use is not restricted to them. Ultraguard is in fact often
used where petroleum distillates such as gasoline and diesel are stored.

Unifire

Fan for **PPV**, available in the power range 4 to 18 hp, in each case having
an engine manufactured by Honda. The fan can be locked in position at
any angle from 10° below horizontal to 20° above it.

Universal Furniture Factory

This factory in south Wales, where furniture sent in "flat packs" from the
Far East was assembled, was the scene in December 2002 of a major fire.
Previously performing strongly, the factory suffered massive indirect losses
from the fire and jobs were lost.

Unlined hose

Canvas and flax are common materials for such hose, although the latter
is susceptible to formation of mildew. Unlined hose is sometimes used on
ships, when it is connected to the **fire main** of the ship. In general, it is
tending to be replaced with "Pin Rack hose", which has equivalent flexibility
but the benefit of a thin plastic lining.

uPVC

Unplasticised PVC, also known as PVCu. Its being unplasticised makes it
rigid. Its uses include window frames in which case, of course, fire codes
and standards apply. Where uPVC is substituted for wood in a window

frame, that is a plus for fire safety as uPVC is much more difficult to ignite that wood is. When a window breaks as a result of a fire, a wooden frame is more likely to be destroyed along with it than a uPVC frame is. Other applications of uPVC are many, and include pipe work.

Urbana, IL

This locality was recently reclassified 2 on the **Public Protection Classification (PPC™)**, scale having previously been class 3. There are only a relatively small number of places in Illinois with a rating this good. A decline in fire insurance premiums of about 3% is expected.

USA, fire death rate

The figure for 2004 was 12.4 deaths per million population. This national figure is straddled by state values for North Dakota and Montana, respectively 12.6 and 11.9 deaths per million population. North Dakota and Montana are of course neighbouring states, each having a border with Canada. Rhode Island, the second most densely populated state, and Wyoming, the least densely populated state, each have death rates well below the national: respectively 4.6 and 2.0 deaths per million population. (See also **District of Columbia, anomaly in the fire death rate of**; **Gulf Coast states, fire death rates of**; **Selected countries, fire death rates in**.)

USA, lightning fires in

Forest fires brought about by lightning occur in the US all the way along the west coast and in the south west and south east. In Washington, California, Nevada and Oregon there are each year of the order of 50 forest fires per million acres (1562 square miles) of susceptible terrain. There are about 20% more in parts of Arizona and Nevada, including the parts of those states which comprise the **Coronado National Forest**. In the south east Florida, Georgia, South Carolina and Alabama have significant numbers of such fires.

US cities, fire hydrants for

Notwithstanding the various designs of hydrant available, some of which, including the **Pacer** and the **Mueller Centurion 250™** have been discussed in this volume, particular US cities have often drawn up their own designs and issued tenders for manufacturers to construct hydrants accordingly. An example is the "Chicago Illinois Spec.", issued in 1916 and known to have been followed at least as recently as 1988. Mueller, whilst developing its own products, including the Centurion as noted, have manufactured

Chicago Illinois Spec. hydrants. Earlier Chicago Illinois Spec. hydrants would simply be embossed with CWW (Chicago Water Works) and would not identify the manufacturer. A fire hydrant once installed has at least as long a life expectancy as a healthy human being and fire hydrant designs also have a very long shelf life. In New York the **O'Brien hydrant**, first acquired by the city in 1902, remains in use.

US Forest Service, attitude of towards piston engine aircraft

This Federal agency has recently followed a policy of not using piston engine aircraft in fire fighting. This precludes, for example, use as an **air tanker** of the **PBY Catalina**. A Catalina operator is not, however, excluded from bidding for a contract awarded by a state fire authority. Note that the **Lockheed Electra** and the **Convair 580** are both turboprops so can be used in USFS work.

USNS[37] *Shughart*

There was a major fire in this vessel in March 2004 after it had arrived in Kuwait and its cargo was being unloaded. Three crew members were treated for smoke inhalation and there was extensive damage to the electrical system of the vessel. Two sister vessels also at Kuwait – USNS *Red Cloud* and USNS *Sisler* – responded to requests for assistance by supplying breathing apparatus. Fire trucks and a fire fighting vessel from the Port of Kuwait also came to the scene. The tugs, which unlike the *Red Cloud* and the *Sisler* had external fire fighting capability, directed water at the Shughart to keep down the temperature of the hull of the vessel whilst fire fighting was taking place on board.

USS *Bennington*

Aircraft carrier upon which there was a fire in May 1954. The fire, the death toll from which was 103, occurred whilst the vessel was at Quonset Point, RI. The cause of the fire was leaked lubricating oil from an aircraft launching device ("catapult"). After repair in New York, the *Bennington* remained in service until decommissioning in 1970, having in the meantime participated in the Vietnam War. It is widely held that the fire on USS *Bennington* is the worst fire in the history of Rhode Island.

USS *Enterprise*

One of a succession of vessels of the US Navy to bear that title (one of the previous ones had been present during the Japanese attack on Pearl Harbour), this vessel was the first nuclear-powered aircraft carrier. In

---

37   United States Naval Ship.

January 1969 she was at Pearl Harbour when she was hit by an accidentally discharged missile from a stationary aircraft. The death toll was 27 (all members of the US Navy) and there were hundreds of injured personnel. Grave though the accident undoubtedly was, what might conceivably have been a further very serious event did not take place: there was no threat to the nuclear power plant therefore no discharge of nuclear material. *Enterprise* re-entered service after repair.

USS *Forrestal*

Aircraft carrier used in the war with Vietnam, the scene of fire and explosion in July 1967. As the planes on the carrier were being prepared for an attack on a North Vietnamese target, there was an electrical surge causing loss of control with the result that one of them released a projectile intended for the enemy. This struck a nearby stationary aircraft which released its fuel and fire resulted. The most serious part of the escalation was activation of a high-explosive bomb composed of TNT and RDX. We might note as a point of interest that normally such a bomb would be initiated with primary explosive, but heat alone can bring about initiation, as happened on the *Forrestal*. The bomb once ignited acted in effect as a primary explosive for other bombs which the *Forrestal* was carrying. The lives of 134 servicemen were lost, the heaviest such loss by the US Navy since the Second World War. The *Forrestal* was taken to the Philippines for repair and remained in service for many more years.

Ussurian pear

Botanical name *Pyrus ussuriensis*, tree native to Asia and introduced into North America where it is classified as a **firewise plant**.

Ultrasound (Ultrasonic) gas detector

This works by detection of the ultrasound emitted when a gas under pressure leaks. Such devices, including the **Gassonic** detector, find application at offshore platforms. They are unique amongst gas detectors in that no contact with the leaking gas itself is required.

UT 512-OPV

Coastguard vessel designed by Rolls Royce Marine AS, having fire fighting capability at **FiFi I** level. It is powered by diesel and is used by the Norwegian coastguard (Kystvakt in the local language), where its duties include tanker assistance. UT 719-R, also of Rolls Royce conception, is a rescue vessel which in the event of the need to evacuate an offshore platform will take up to 300 evacuees to safety. It doubles up as a fire fighting vessel with FiFi rating depending on the buyer's specification. As yet "on the drawing board", the next vessel in the UT series will be the *UT 776*, which will be available in a configuration for fire fighting only. From the same stable is the *FD Incredible*, a platform supply vessel delivered in March 2007 for use in the North Sea having FiFi I classification. The **Malila** (UT 719-2) was also designed by Rolls Royce Marine as was *Geonisio Barroso* (UT 722).

Utorongun Gas Plant

Natural gas facility in Nigeria operated by Shell by means of which non-associated gas (that is gas other than that occurring with crude oil) is released into the country's gas reticulation system for supply to homes and industry. In October 2007 two fires occurred at the facility and the shutdown necessitated significantly affected gas supply.

UV/IR flame detector

This has one photoelectric tube for UV and another for IR and activates an alarm when a current is present in *both*. The **SenseTronic® T-229/2** has a UV detector and an IR detector which act both independently and as a UV/IR detector. The UV/IR flame detector might be of advantage where there is the possibility of unnecessary activation from ambient radiation at one extreme or the other of the visible region, e.g. where welding is taking place.

UWB Wireless Fire Fighter Locator

As yet under development, this uses the principle of the **Pak Tracker™** and other transmitting **PASS** devices, whereby a radio signal is emitted from a fire fighter's apparel. Instead of simply a hand-held receiver, however, there will be two or more receivers installed on fire trucks outside the building. It is believed that locations of several fire fighters in a building can be monitored continually to within a metre by this means.

Vancouver International Airport, fire in an Airbus at

Wiring in the in-flight entertainment system is believed to have been the origin of a fire on an A330 aircraft stationary at Vancouver International Airport in January 2002. There were no passengers on board. The entertainment system was turned off and a halon extinguisher was applied to the fire. The aircraft was able to take off on time on the flight for which it was being prepared at the time of the incident. Investigation focused on the (entertainment) system management unit (SMU). A fan assembly from within the unit was found on examination to have become detached. The fan had "burn patterns", as had a circuit board close to it. Other evidence of considerable heat generation was melted solder. It was noted that the turning off of the entertainment system had not taken immediate effect, as when it ceases to receive power from the aircraft's own supply it draws on batteries for two minutes whilst the software shuts down.

Vari-Nozzle

This nozzle attributable to Akron[38] can be used as an example of two of the principles of nozzles given in other entries. It is a **fixed gallonage nozzle** delivering 95 gpm at a pressure of 100 psi. It can issue water as a straight stream or as a fog, the g.p.m. not being significantly affected. Its construction material is brass. Aluminium is at least as common in the construction of nozzles for fire fighting, as with the **Pokador 500/750** range. Some are entirely non-metallic.

Vegetation, spark ignition of

There are many documented cases of this, where electrical conductors pass over or close to vegetation and where power tools are being used close to vegetation. For example, in May 2007 in Orange County, CA fence

---

38 Akron Brass Company, Ohio, formed about 90 years ago as a spin-off from B.F. Goodrich to make, from brass, fittings for the rubber products of the latter.

installation was taking place when a spark from a metal cutter ignited vegetation debris. Strong winds aided propagation and the fire developed into a major one, which was fought partly from the air. (See also **Yorba Linda, fire 2005**.)

Vehicle fires, frequency and consequences of

In 2004, US fire departments attended in excess of a quarter of a million fires which had started in vehicles whilst they were being driven. Such fires caused 520 deaths and $\approx$ 1300 injuries. Although a vehicle fire can be caused by collision and fuel tank breakage, this is not the most common cause. Mechanical and electrical faults account for about two thirds of the vehicle fires in the US.

Vehicle-mounted pump

Most pumps mounted on an appliance for water direction at a fire, and all newly made ones in the US, are of the centrifugal type. This has a rotating impeller and the release rate of water is dependent upon this as well as on the pressure. **Waterous** produce many of the vehicle-mounted pumps used by the fire departments of the US (and elsewhere) and performance data for its "no frills" CS model are delivery of 1250 gpm at a pressure of 150 psi. The Waterous range extends to release rates of 3000 gpm and these are often referred to as **midship pumps**, as are comparable pumps from Hale and from Darley. In the CS model the impeller is made of brass. The approximate counterpart in the UK of the Waterous series of pumps is the Godiva series, made in the English Midlands. These are supplied to fire services in very many countries. The **Godiva World Series** is the most recent product from this manufacturer.

VektorFlo®

Hydraulic oil, petroleum based and having a flash point of 210°C. It is a product of Vektek whose HQ is in Emporia, KS. One of the several other hydraulic fluids for which VektorFlo® can, according to its manufacturer, be substituted is **Tellus 32**. One of the **Texaco Rando® HDZ** series is also a suitable substitute, as is one of the **Nuto H** range.

Ventilator, fire protection of

A conventional ventilator consisting of a series of parallel metal slats can make for difficulty during a fire in two ways. The obvious one is that it ventilates the fire, aiding its development. The less obvious one is particularly relevant if such a ventilator grid communicates not with the outside but between two rooms, as is often the case when a building has fan-forced ventilation, which a door lacking a ventilator grid would obstruct. In that event, unless the slats are so close together as to quench a flame incident upon them (in which case the ventilator is acting like the miner's safety lamp invented by Sir Humphry Davy) the ventilator provides a path for propagation. Ventilators are available in which the slats are coated with an intumescent material, which will swell and close off the spaces. (*See also* **Astro Damper**.)

Ventry®

Fan suitable for **PPV**, available in blade sizes 20 and 24 inch and having (like so many other such fans) Honda engines. Air movement rates are up to about 20 000 ft^3 per minute. A notable feature of fans in this range is that the blades are made from wood which is reinforced with fibre glass and Kevlar®. There is the obvious advantage that such a material is much lighter even than aluminium, providing a greater rotation speed per unit power delivered at the shaft. A second advantage is that for a particular fan the wooden blades can be carved into a configuration to meet a performance target. Also using non-metallic blades is the **Fantraxx** series.

Vertol 107

Twin-rotor helicopter, a Boeing product, having been widely applied to aerial fire fighting. If used in extinguishment, such a helicopter can carry 500 gallons of water in a **Bambi Bucket**® or equivalent. The type is however also used in **controlled burning** and was in fact so used during the severe forest fires in California in October 2007.

Viking fire hose

This has the nature of two layers of synthetic rubber with a woven layer in between, referred to by the supplier Macron as a "unified lining and cover". **Reel-Lite 800** also has this sort of structure. The word "unified" signifies a point of major importance: integrally constructed fire hose not containing an adhesive has a longer life expectancy and higher reliability. Warranties of five years' duration are available with such hose.

Vinylflow

Proprietary fire hose for low-pressure conditions, for use only up to about 70 psi. It has an inner layer of PVC covered with woven polyester, the outside also being PVC. Thus it has the structure of a typical **booster fire hose** but cannot be used as such because of the low pressure rating.

Virginia Beach

Close to the border between Virginia and North Carolina, this coastal locality is a holiday destination. Sprinkler systems became mandatory in hotels at Virginia Beach in 1997 and there was some delay in the response. Senator K. Stolle recommended an extension on the date by which sprinklers had to be in place. Hotel owners pointed out that the most recent vacation season had been sluggish and that money for the installation of sprinklers – about US$100000 per hotel at that time – was not available. They also appealed for low-interest loans or tax relief on the costs of sprinkler installation.

Volkan

Small (≈ 100 employees) manufacturer of fire fighting equipment, based in Turkey, having been in the business since 1974. Amongst its products is a TL vehicle with a ladder reach of 32 m and a water tank capable of holding in excess of 1000 gallons. The pump (up to 800 gpm) and **monitor** are both of Volkan's own manufacture.

Volunteer Fire Assistance (VAS) Program

It is recognised that vast areas of rural USA are protected from fire primarily by volunteer fire fighters. In such areas volunteer fire crews can receive financial grants from Federal sources as well as technical advice and support and help with training. For the purpose of the program a "rural community" has a population of 10000 or less.

Volunteer fire fighters, financial value of the work of

If the work done by volunteer fire fighters in rural areas of the US were all done by "career fire fighters", the cost would be over US$30 billion per year. Yet there are many US volunteer fire services which operate on a budget not exceeding US$10000 per year. (See *also* **Kansas, forest fire protection in**.)

Vulcan RF™

A **monitor** currently leading its competitors in one important respect: it can be remotely operated from a distance of ¼ mile from the monitor itself, meaning that the fire fighter can be that distance plus the "throw" of the monitor from the place of direction of the water. It is manufactured by Darley in Illinois, who have been making fire fighting equipment since 1908 and manufacture a range of monitors including the **Mercury**™ and also hose made of **nitrile**.

Wadi el Natrun Prison

This prison near Cairo was the scene in 2007 of an electrical fire which killed three inmates and hospitalised 18 more. The fire originated at a ceiling fan.

Wallenius Lines

Swedish operator of about 180 vessels including many car and truck ferries. It has retrofitted 27 of its vessels with **Hi-Fog®** systems for fire extinguishment, these being directed at vulnerable parts of the respective vessels including the engine rooms. The vessels did not have to go into dry dock for the retrofitting: in some cases it was carried out in foreign ports.

Wanganui

This town in New Zealand's North Island was recently the scene of a fire at a paper recycling factory. The building is at the waterfront and the wharf was threatened by the fire.

WASP

Wide-range accurate sprinkler proportioner. Such a device will often use AR-FFFP. There is another difference, however, between a conventional **sprinkler head** and one having a WASP. In the former water exiting is deflected by impact with a horizontal disc and the rest is left to gravity. A WASP will usually have a conical spray profile. The cone angle and the base length – a measure of the area which the spray will cover in its fire extinguishment function – is controllable by the orifice diameter and the water pressure. Such information is known very precisely for several commercially available WASP systems, making protective planning possible.

Water, for extinguishment, straining of

Where water for fire fighting is taken from a pond or river, initially to a **water tender**, straining will be necessary to remove leaves and other debris.

Water, velocity of when exiting a fire hose

The simplest is the barrel strainer, a cylinder with holes drilled in its walls. Water for application to a fire is freed of trash on passing through the holes. Aluminium is a common choice of material for such a device. A strainer which is actually at the source is called a floating strainer. The water to it is drawn to a filtering screen, which can be positioned at a depth below the matter on the surface of the water, such as leaves, but above particles originating at the bed, such as sand and silt.

Water, velocity of when exiting a fire hose

Application of Bernoulli's equation gives:

$$v = \sqrt{(2\Delta P/\sigma)}$$

where v = exit velocity (m s^{-1}), ΔP = gauge pressure of the water, i.e. the amount by which the water pressure exceeds atmospheric (Nm^{-2}) and σ = density of the water (1000 kgm^{-3}: more precise values than this are unnecessary in such calculations).

The treatment makes the assumption that the velocity of the water within the hose is negligible in comparison with that on exit. A value of ΔP of 10 bar would be somewhere along the very wide range of water pressures used in fire fighting practice, and this gives:

$$v = 45 \text{ m s}^{-1}$$

How powerful this is can be perceived by calculation of how high water released at such a speed would rise against gravity if the exit from the hose were pointed vertically upwards. This height (h, unit m) is given by:

$h = \sqrt{(v^2/2g)}$, where g is the acceleration due to gravity, giving:

$$\underline{h = 10 \text{ m (33 feet)}}.$$

This height is a criterion for classification of a vessel as **FiFi I**, **FiFi II** or **FiFi III**.

Water curtain

Spray (fog) issuing from a nozzle at a wide angle, possibly 160°. Often used in high-rise applications, such a curtain can have a diameter in excess of 10 m. Nozzles are available which provide for such water release or for a straight stream depending on adjustment. A narrower angle at an intermediate nozzle setting provides a "cone" of water.

Water ladder

A.k.a. a water tender ladder, fire appliance with pumping and water storage facilities carrying **ground ladders** or roof ladders only, that is no aerial ladder.

Water mist fire extinguisher

Water mist for extinguishing purposes needs to have a size distribution such that almost all (>99%) of the droplets are of diameter less than 1 mm. The discharge pressure will be in excess of 10 bar. There has been a revival of activity in pure water as an extinguishing agent, partly because of its environmental benefits, including the fact that it obviously does not threaten the ozone layer, nor does it produce any toxic products. Much R&D has taken place and positive results in the extinguishment of many types of fires, including electrical ones, have been reported.

"Water only" fire extinguisher

Enormous benefits have resulted from foams, wetting agents and the like in the extinguishment of fires with water as the primary extinguishing agent. Even so there is still an important place for water alone and fire extinguishers containing water only are available, being used on Class A fires in particular. For example, Chubb manufacture such an extinguisher which has a capacity of 9 litres of water and discharges in just under one minute. This is achieved by having the water at high pressure in phase equilibrium with the small amount of vapour in the head space, and design conditions relating to the extinguisher wall thickness apply. If hypothetically such an extinguisher were to fail and its water contents be released, the result would be a boiling liquid expanding vapour explosion (BLEVE).

Waterous

Manufacturer of fire fighting equipment since 1886, with its HQ in St. Paul, MN. Manual water pumps for fire fighting were replaced with ones driven by steam and later with pumps using an internal combustion cycle. The first such pump to be powered with gasoline was made by Waterous in 1898. Its recent products include the **Advantus**™ foam proportioner. Waterous are also the manufacturer of the **Pacer**. The company also produce various models of **vehicle-mounted pump**.

Water tender

Mobile fire fighting unit for carrying water to where it is required for extinguishing a fire. The source might be a hydrant or, in rural areas, a lake or pond. A water tender will have a capacity of 1000 US gallons or very often more. A tender can empty its contents into a **dump tank** and then return to the water source for replenishment and in that way reserves are built up for use by the **attack vehicle**. An appliance might carry foam concentrate (or several thereof) only, for combination with water from a hydrant by means of a proportioner at the scene of the fire, in which case the term *foam tender* applies.

Water thief

Device whereby a single line of water from a nozzle can be split into two or more such lines. In its simplest form it is just a T-piece or Y-piece and might be constructed of plastic. Basic and less expensive forms use screw thread attachments, but more advanced designs will incorporate **Storz coupling**.

Water tower truck

Fire appliance from which water can be directed at a fire from well above street height. The term is not *usually* used synonymously with "aerial", in which the water is directed by a fire fighter from a ladder. It might be used synonymously with **hydraulic ladder** or hydraulic platform. In general usage "water tower" means water stored at elevation so that it flows in response to gravity when required. This is not true of a water tower truck, in which water is received from a hydrant or **water tender** via a pump.

Watertown, WI

The scene in 2005 of a fire at a tyre recycling plant. There were more than a million tyres at the plant, well in excess of the number it was licensed to hold. The plant is some distance from water hydrants. Over quite a wide area outdoor recreational activities were put on hold until the smoke had dispersed.

Westbourne Park, London

The scene in January 2006 of a fire at a bus depot, likely to go on record as one of the worst "vehicular fires" ever in terms of financial loss. There are reports of explosions during the fire, which almost certainly were due to fuel tank rupture. There was also much smoke, which can be attributed to the combustion of polymer materials in the buses and perhaps to the tyres.

West Columbia, SC, Fire Department

Recipient in 2007 of a new aerial fire fighting appliance on a Pierce cab and chassis structure. Delivered incomplete to a local builder of fire apparatus, it was fitted with equipment including a 2000 g.p.m. **Waterous** pump and a 300 gallon tank. It is fairly common in the US and the UK as well as in Japan for an appliance to be so built and the point has been made in another entry in this volume that a particular type of fire truck might exist with different badges on the front of the cab. The cab/chassis are one component only of an advanced piece of apparatus and will be sourced in the same way as ladders, water tanks and pumps. That is why to talk of, for example, a "Mercedes fire engine" is not rigorous. The celebrated "Green Goddess" would on that basis be a "Bedford fire engine" but is seldom if ever referred to as such. (*See also* **Angloco**.)

Western wheatgrass

Botanical name *Pascopyrum smithii*, grass having a good resistance to fire spread by reason of its thin distribution, which makes for a low fire load. It is possible to protect a home in the event of forest fire by planting around it seeds from a suitable mix of "firewise" grasses. This is being done for example in New Mexico, and western wheatgrass features in recommended blends of seeds for this operation.

Wet barrel fire hydrants

Examples of fire hydrants described in this volume under their trade names have tended to be of the dry barrel type, that is provision is made for removal of water from the barrel when use of the hydrant ceases. The reason for this is an obvious one: were water in the barrel to freeze it would render the hydrant inoperative. Equally obviously this is not an issue where the climate is such that such water will not freeze and in states of the US, including Florida and California, wet barrel hydrants, in which the water in the barrel is retained on cessation of operation, are in use. Notwithstanding the use of dry barrel hydrants in colder locations, freezing of hydrants does sometimes occur in which case a **fire hydrant thawing unit** is brought into use.

"Wet chemical" extinguishing agents

These contain solutions of potassium citrate ($K_3C_6H_5O_7.H_2O$) and/or potassium acetate (CH_3COOK) and have been developed for Class F fires. The organic moiety of the salt can react with fats and oils to form a soapy layer, the formation of which leads to extinguishment. Having regard to the fact that such an extinguishing agent is "wet", an additional ingredient might be an **AFFF** or a **FFFP**. The **Jewel 6L** is an example of a wet chemical fire extinguisher.

Wet type sprinkler

Sprinkler in which water is present in the pipe above the **sprinkler head** at all times, making for rapid release when the sprinkler is engaged. This is the most common type of sprinkler. As with hydrants, there can be difficulties with sprinklers if they are in an environment where the temperature can drop to 0°C or below. A wet type is not suitable under these conditions.

"Wet wing" tankage

Term applied to a fuel tank in an aircraft which is integral, that is not an installation that can be removed and replaced. Many commercial passenger jets have an integral tank in each wing as well as a non-integral one attached to the fuselage. In 1984 there was destruction of a Boeing 737 (although no deaths or injuries) when an engine component became detached and punctured an integral fuel tank in the wing to which it was attached. (*See also* **PR1005L**.)

"Whisper" fan

Ventilation units for **PPV** create enough noise to necessitate hearing protection, an exception being the "Whisper" PPV fan. Like the **Unifire**, the "Whisper" can be positioned for use at any angle from 10° below horizontal to 20° above it. Its design and development were largely to meet the needs of fire services in the UK, who tend to specify this angle range for fans used in PPV. The "Whisper" has Honda engines.

Whittlesey

This East Anglian town was the scene in 2004 of a factory fire in which two employees died. The fire, at the premises of Anvil Alloys International, is believed to have been caused initially by a spark generated during a metal grinding operation. The factory contained many cylinders of gas, some of which exploded. It is alleged that the two victims had been trapped behind a stuck fire door.

Wildwood, NJ

The scene in December 2005 of a fire of **alarm number** 9. It occurred at a motel, the second and third rows of which were burning strongly by the time the FD arrived. There were no persons inside the motel at the time. The ground floor was occupied by retail outlets, which were totally destroyed.

Willows, CA

Often viewed as the birth place of aerial fire fighting, the scene in August 1955 of departure of an adapted Boeing aircraft for a water drop on to a fire in the Mendocino National Forest. The aircraft had a water holding capacity of 170 gallons, less than half that of the **Dromadear**, which is currently very much at the small end of **air tanker** sizes.

Wired glass

The *raison d'etre* of wired glass is that the wire holds the glass together in a fire, whereas without the wire support it would have shattered. When glass safety is considered, both fire resistance and impact resistance are relevant. For a period in the US there was a move away from wired glass in schools and the like because of poor impact resistance resulting in laceration injuries. Unless the glass is suitably coated to make it a **low-e glass**, it is likely to have an emissivity towards a wide range of wavelengths. This can be mitigated by polishing the glass surface in manufacture. (*See also* **Pyrostem**.)

Woodside, NY

This area within the New York Borough of Queens was the scene on 27 December 2007 of a fire in a historic church. St Paul's Episcopal Church in Woodside, built in 1874, experienced severe damage including the loss of valuable stained glass windows.

Woolfardisworthy Sports and Community Centre

This facility in north Devon, England was destroyed in a fire in December 2003. Wood featured very centrally in the construction and this aided fire spread, as did strong winds.

Wyld, Obadiah

Originator of the first patent for a substance to retard fires in fabrics. This was in 1735 and the retarding substance included **borax**, which, in the 21st century, continues to find such application on a grand scale.

Wylie, TX

The **Public Protection Classification (PPC™)** rating of this city is "split". Properties in it within 1000 feet of a hydrant and within 5 miles of a fire station are class 1: others are class 9.

Wynnewood refinery, OK

The scene in early 2008 of a propane fire. Closure of highways near the refinery was necessitated but there were no injuries. The incident was one of a succession at that refinery over a period of about two years.

X³ ® Technology

Term applied by **MSA** to a gas-detecting device containing catalytic, electrochemical and infrared sensors in one detector unit.

XP95

Range of smoke detectors manufactured by Apollo Fire Detectors in the UK. They are of the ionising type and therefore contain ^{241}Am. XP95 detectors incorporate intrinsic *safety* in the following regard. The level of electrical power which such a detector draws is too low to constitute an ignition source if a flammable gas or vapour enters the detector. This is preferable to the alternative of having a flameproof housing for a detector, the reliability of which will never be 100%.

Y

Yaoundé train explosion

A train was hauling fuel oil through Yaoundé, the capital of Cameroon, in February 1998 when it derailed close to a major rail intersection. Tank cars broke open and their flammable contents leaked out. Even so, at that stage there were no deaths or injuries: that came later when persons tried to pilfer the leaked fuel. One such was smoking a cigarette which ignited the vapours and began a sequence of events which included a flash fire and a fireball. Over 120 people were killed (the exact number is not known), and at least as many received serious burns. There was also destruction of slum housing nearby. Railway infrastructure was poor, partly because of a corrupt regime in which money for public services is misappropriated, and the consensus is that this was the cause of the accident. France sent financial help and several eminent French burns specialists went to Cameroon to help the injured. It is interesting to speculate on what risk assessment if any had been done on the passage of a train bearing tankcars of hydrocarbon through such a densely populated place.

Yekaterinburg

This town in the Urals was the scene of a fire in a furniture factory in which there were 12 deaths. Evacuees jumped through windows, there being no proper fire escape.

Yellow stonecrop (a.k.a. lanceleaf stonecrop)

Botanical name *Sedum lanceolatum*. This **firewise plant** grows to about 30 cm in height and has leaves of 2 to 3 cm length and yellow flowers. It occurs on the west side of North America from Alaska down to California, and also in Colorado, Nebraska and North Dakota.

Yogyakarta airport, Indonesia

The scene in March 2007 of the explosion of a Boeing 737 aircraft on landing. The death toll was 22. It is believed that the front wheel of the aircraft was on fire before the landing. An investigation concluded that the response by fire fighters had been too slow and that foam extinguishing agent had been unavailable.

Yorba Linda, fire 2005

It appears the Independence Day celebrations having continued into 5 July resulted in this major fire in Yorba Linda,[39] which was caused by fireworks. The fire was at a site distant from housing and no buildings were lost. There were, however, eight fire fighter injuries. The isolation of the scene of the fire caused difficulties with fire appliance entry, made worse by recent rains. Amongst the fire departments participating were some from Orange County.

York, household waste fire

Household waste has a calorific value such that a tonne of it releases as much heat as a barrel of crude oil. Some components such as paper and cardboard are very easy to ignite and therefore a temptation to arsonists. This happened in the historic English town of York in 2004, when a fire was started in a pile of previously collected household waste.

Yorkville

District of Toronto, recently the scene of a fire at a restaurant. Initially classified as a two alarm fire, it had risen to a five alarm one by about an hour after the fire department were first called and 130 fire fighters attended.

---

39   This town is most noted as the birthplace of President Richard M. Nixon.

Z

Z8

ARFF vehicle manufactured in Germany by Ziegler. It has four axles and a **turret** on the bumper and another on the roof. There are five, in predominantly yellow livery, in service at Zurich Airport. The Z8 is also to be found at Kuala Lumpur International Airport and at Mactan (Cebu) Airport in the Philippines.

Zanderij Airport

This airport serves the city of Paramaribo, capital of the tiny South American country Suriname (or Surinam). In 1999 the Dutch airline KLM, the only major international operator to use the airport, expressed concerns at substandard fire protection at the airport, in particular the maximum water discharge capability which was below that required for Boeing 747 movements. Even the nominal water capacity of the airport could not be relied upon as the fire trucks at the airport were in poor condition.[40] In the event two fire trucks were loaned from the Netherlands until such time at new ones, enabling the requirements of a 747 to be met, could be acquired by the airport itself.

ZB-223

Smoke suppressant, manufactured by Great Lakes, having zinc borate as its primary constituent. It is often incorporated into polymers, being most suitable for those which experience relatively low (< 200°C) temperatures during moulding. Sister products are ZB-460 and ZB-467. The former can withstand, without loss of smoke suppression capability, temperatures up to about 500°C, whereas the latter, which finds application to nylon in particular, is limited to about 300°C.

---

40   That was not always the case. What was then a state-of-the-art fire truck by Daf was supplied to the airport in the mid-1960s.

Zeroflame Varnish

Flame-retardant coating material for wood surfaces. In its application a colourless water-based layer, containing a substance capable of **intumescence**, is first applied and once it has dried a solvent-based layer is added. From the same manufacturer (in Wichita, KS) comes the **FlameOut™** kitchen towel.

Zeta optical smoke detector

This is of the type where in the absence of smoke none of the infrared light from the built-in source reaches the photodiode: it is diverted to it by scattering when smoke enters the detector. It can operate on a 9 V battery.

Z Gard® S

Gas detector from **MSA**, for gases including carbon monoxide. It is available with a solid state or with an electrochemical detector.

Zil (Zavod Imeni Likhacheva)

Russian manufacturer of trucks, some of which were adapted for fire fighting use, in particular the Zil-130/131 series "pumpers", which are still in service in places, including Moscow. In spite of its relatively short wheelbase the Zil-130/131 has been made into an "aerial" and in this form has entered service in Cuba. One of Zil's current models is the **Fire bluster**.

Zinc hydroxystannate (ZHS)

Chemical formula $ZnSn(OH)_6$, a widely used fire-retardant substance when it also acts as a **smoke suppressant**. Excellent retardancy results have been obtained with ZHS in applications including PVC cables, polyester resins and nylon. Zinc stannate, $ZnSnO_3$, usual abbreviation ZS, also finds application as a fire retardant. Both ZHS and ZS work as fire retardants by promoting char formation to the exclusion of pyrolysis. The importance of ZHS and ZS and their scale of production is an incentive for the recycling of tin.

Zinc oxide (ZnO)

Like **tin(IV) oxide** an n-type semiconductor having found application to gas sensing. Doping is possible with aluminium, chromium or indium.

Zirpro®

Fire-retardant material for wool, developed by the International Wool Secretariat (IWS) in collaboration with chemical manufacturers. It comprises zirconium and titanium compounds. The term is sometimes broadened, and "Zirpro wool" means wool treated with zirconium compounds. Being composed of keratin which is a protein, wool burns quite differently from cotton and synthetic fibres in that ignition is prevented from fully developing by formation of fluid. This effect, which is as if the wool had a built-in intumescent retardant, is termed "foaming" in some coverages of the subject. The role of zirconium and titanium compounds is to promote the "foam" formation. Zirconium compounds so applied include the hexafluoride and the acetate. **Titanium chloride** alone is an effective flame retardant.

Zotek®F

Polyvinylidene fluoride polymer, material, closely resembling **Kynar®**, used in aircraft construction as thermal and acoustic insulation, also in seating and trim. Its density can be varied in manufacture and specified by a purchaser from 30 kgm^{-3} upwards.

Postscript

It is recorded on the web page of Dr Vytenis (Vyto) Babrauskas that he was "the first person ever to be awarded a PhD in Fire Protection Engineering". That was in 1976 and Dr Babrauskas, an associate and friend of the present author, is now one of the leaders of his profession. Dr Babrauskas can therefore be said to have been one of the pioneers of the scientific discipline of fire protection engineering. The point being made is that we are still in a sense in the first generation of fire protection engineers trained as such. By contrast, the very first chemical engineering graduates departed MIT not many years after the First World War so the commencement of the profession is outside living memory.

Over the 35 years or so of its existence as a distinct profession, fire protection engineering has expanded to such an extent that many universities offer Baccalaureate level courses in it. Students enrolling in such courses are committing themselves at the very beginning of their working lives to a particular profession in much the same way that medical students and law students do. Another entry to the fire protection engineering profession is a science degree followed by a Masters program in fire protection engineering. In either case, for UK students the goal will be corporate membership of the Institution of Fire Engineers with associated Chartered Engineer status. As with any other profession, a newly qualified fire protection engineer will not see him/herself as having "arrived" but will want to participate in approved professional development activity.

Accordingly, it is intended that this book will be of value to those training in fire protection engineering and to those already practising it. Conceivably homeowners (or planners of newly constructed homes) who want to take a DIY approach to fire protection will benefit from the book. Finally, the author earnestly hopes that there will be those who read the book simply because they find it interesting. If such interest is taken, fire protection engineering can be seen not only as a profession but also as a discipline having intrinsic intellectual appeal and the rather unnatural distinction between a "pure" and an "applied" subject is blurred.

10